AVALANCHES
AND SNOW SAFETY

AVALANCHES
AND SNOW SAFETY

Colin Fraser

JOHN MURRAY

Printed in Great Britain by
Richard Clay (The Chaucer Press), Ltd.,
Bungay, Suffolk
0 7195 3479 8

*To the Men of the Parsenndienst
in Gratitude and Affection
and to the Memory of Christian Jost
Guide and Mentor*

Foreword

by Chris Bonington

Avalanches are one of the main hazards facing the mountaineer or ski alpinist, and I, and expeditions I have led, have frequently had to contend with them.

More often than not, they claim the unwary and unprepared, but even people of enormous mountain experience may be trapped by them. Just over a year ago, Dougal Haston, a close friend with whom I undertook some of my most demanding of mountaineering endeavours, was killed in one while skiing in Switzerland.

People continue to lose their lives in avalanches year after year, and it is only through a more widespread understanding of the avalanche phenomenon that the death roll can be reduced. For this reason, I particularly welcome this informative book. When the earlier edition was published in 1966, it was recognized as an authoritative work, and in serious mountaineering and skiing circles, it quickly achieved the status of 'essential reading'. Since his period of work with the Parsenn rescue organization and with the Swiss Federal Institute for Snow and Avalanche Research—during which he prepared the earlier edition—Colin Fraser has maintained his contacts with other avalanche specialists and has frequently worked as an avalanche consultant. For this reason he has now been able to bring the book fully up to date, adding material on recent research results and rescue innovations.

The book is aimed at the mountaineer and skier rather than at the snow scientist and it covers the subject in detail. It is readable and lucid and anyone reading it cannot fail to acquire a fuller understanding of the snow cover, of why avalanches occur, of how to avoid or reduce the risk of being caught in one, of how to rescue victims, and so on. At the same time, the reader will acquire a healthier respect for the menace that may lurk in a snow-covered slope, for the technical subject matter is illustrated by many factual and dramatic accounts of actual accidents.

All of us who go into the mountains do so in the knowledge that we are exposing ourselves to risk. But at least as far as the risk posed by avalanches is concerned, we can arm ourselves and reduce it by reading and absorbing the material in this book.

Contents

Illustrations

FIGURES IN TEXT

Author's Note and Acknowledgements

This work originally appeared as *The Avalanche Enigma* in 1966. It was published under that title in Britain and the U.S., and German and Italian versions followed. This new edition has been entirely updated in the light of recent knowledge gained from research, and also to cover the latest developments in the field of avalanche rescue. The title has been changed in order to distinguish the character of the book more clearly from the several novels that have appeared in very recent years and which have used avalanches as a dramatic backdrop.

In order to write the original book, I worked for most of three winters as a patrolman with the Parsenndienst, the rescue and safety organization covering the Parsenn ski area at Davos/Klosters. I believe I was the first Anglo-Saxon ever to work as a patrolman in the Alps, and I shall always be grateful to the men of the Parsenndienst for the way they accepted me into their midst and trained me in their skills, even to the point where I was regularly bringing down injured skiers on sledges, unaided, and participating as a full member of patrols sent to control avalanches with explosives. I was very fortunate to work under the orders of Christian Jost in his last years as head of the Parsenndienst. His support for my project did not take an effusive form, for he was not that sort of man. But the form it did take was far more useful in that he systematically, like the good schoolmaster he had been, imparted as much to me of his vast experience and wisdom as words alone could convey. I cherish the memory of the many hours spent with him, and hence the dedication of this book.

While working with the Parsenndienst, I was also granted free access to the Swiss Federal Institute for Snow and Avalanche Research. For this, I have to thank its Director, Dr. Marcel de Quervain. The Institute staff were most cooperative and helpful, allowing me to participate in their work, in so far as possible, and answering numerous questions. But, in particular, I came under the wing of André Roch. At that time, Roch was the head of one of the

1

Institute's research sections, but he has since retired to his native Geneva. A famous alpinist—five times in the Himalayas including the Swiss 1952 Everest expedition—a painter, the author of several books, and a noted avalanche researcher and expert, André Roch was unbelievably generous to me. Without his knowledge, advice and friendship the first version, and hence this edition also, of this book could not have been written.

In researching the first version of this book, I received kindness and help wherever I went. The Royal Geographical Society and the Alpine Club allowed me to use their libraries; busy people like Dr. Gino Eigenmann, Professor Haefeli and Dr. Hossli gave generously of their time to talk to me; people like Gerhard Freissegger and others who had suffered terribly in avalanches relived their experiences for me; many great international avalanche specialists answered letters in detail, and so on. There are too many debts of gratitude to be able to identify them all individually.

Between the appearance of the first version of the book and this revised edition I maintained my profound interest in the subject and widened my experience by running numerous training courses, giving lectures, helping to make documentary films on avalanches, and acting as a consultant to ski-areas in need of advice, or of training, or of blasting operations to control avalanches. So when my publishers suggested bringing out this new edition, I was glad of the opportunity.

Once again, however, I have incurred debts of gratitude to a number of people. In particular, I must thank my friend Walter Good of the Swiss Federal Institute for Snow and Avalanche Research. He provided me with a vast amount of updated information about his work and that of the Institute in general. My special thanks also go to Ruth Eigenmann of the International 'Vanni Eigenmann' Foundation for her rapid and full response to my request for information. Many others among my friends and acquaintances in the avalanche world have replied to letters or passed me valuable documentation. They are too numerous to mention singly, but they are not forgotten, and I am grateful for their help.

Two points about this book require some explanation. I have included in it many factual accounts of avalanche accidents in order to illustrate technical points. Some of these accounts are perhaps horrifying, but it is far from my intention to create unnecessary alarm. On the other hand, I do hope to engender a greater respect for, and knowledge of, snow and avalanches than at present exists in many quarters. My greatest pleasure, in the years after the appearance of the first edition, was when readers told me how much their understanding of snow and avalanches had been increased by it and how much more pleasure and safety they found in the mountains as a result. And when a reader wrote from New Zealand to say that the advice in the book had saved his life, my gratification was complete. I hope this new edition will be equally useful.

Finally, a point about Davos: this famous resort and the magnificent Parsenn skiing area feature regularly throughout the book, but this is merely because it was the obvious place for me to go when gaining further knowledge and material. It is the home of the Federal Institute for Snow and Avalanche Research, and also of the Parsenndienst; and it must not be thought—just because I mention so many avalanche accidents there—that it is any more dangerous than other places. If anything, the contrary is true, and attention is focused on Davos owing to the excellence of the safety measures and the importance of the research carried out there.

Rocca Sinibalda, Rieti, Italy

I
Avalanches in the Past

Avalanches, which must be ranked with other great destructive phenomena like earthquakes and hurricanes, have terrorized mountain people for centuries, especially after great snow-storms like those of 1951. Fortunately, storms of such magnitude occur infrequently, but this is one of the reasons for the death and destruction that they still bring. In the past, the avalanches that killed a family here and there or destroyed the odd house each winter were considered to be among the unavoidable and not very grave hazards of living in the mountains. After one of the less frequent major disasters, which had perhaps killed several hundred people—and left no doubt as to the gravity of the hazard—the survivors considered themselves safe for a number of years to come, at least by the law of averages. They therefore shrugged the catastrophe off fatalistically and rebuilt their houses where they had stood before.

It is a quite recent idea that avalanches are not entirely irresistible, indeed that measures can be taken against them, and it was this realization which sparked off modern research into their complex nature, on the basis that knowledge of an enemy is the best defence. However, our forbears can well be excused their fatalistic attitude when one considers the destructive force of avalanches and their very high speed of movement. Airborne-powder avalanches (see photographs 1a, b, c), in which clouds of snow lift from the slope and flow through the air, can reach speeds in excess of 200 m.p.h.

The so-called Great Glärnisch avalanche is a classic example of an avalanche that moved at very high velocity. It occured on March 6th, 1898 at 11.20 a.m. and was observed and reported on by Dr. Beuss, the pastor of a village near Glarus, Switzerland. He was conducting a funeral at the time. The enormous avalanche broke away near the summit of the Vorderglärnisch. The snow-cloud was vast with the sun lighting up the glinting particles near the fringe of the swirling mass. The centre was like a gigantic waterfall, while the turbulent clouds spread out either side as if wanting to start their own streams. The swirling of the snow was creating ever changing ring patterns illuminated by the sun. The ground shook as in an earthquake, and the noise as the snow crashed down over the rocks into the valley was overwhelming.

The noise took on the tone of distant thunder as the avalanche shot across the valley floor, hit the slope of the Buttenwand opposite and then surged upward many hundred feet before recoiling in a curve through the air. The sky went dark as the snow-cloud blotted out the valley and the surrounding peaks. Although the people of the village were well used to avalanches, Dr. Beuss wrote that 'the immensity of this one set the women screaming and running for their houses, while many of the men were noticeably pale'.

The vertical drop of the avalanche was at least 5,750 feet and it ran about $4\frac{1}{3}$ miles, which included $1\frac{1}{2}$ miles of level valley floor, in a little over a minute. It was another 7 minutes before all the snow had settled. The avalanche had travelled at $3\frac{3}{4}$ miles per minute, 225 m.p.h., a speed which Dr. Beuss described as being 'more than four times faster than the fastest English train'.

Modern measurement has shown that speeds of 110–180 m.p.h. can generally be expected from airborne-powder avalanches, while especially large ones may travel at over 220 m.p.h., so Dr. Beuss was surprisingly accurate in his estimate.

There is a blast of wind associated with airborne-powder avalanches which can lay waste acres of full-grown forest in a few seconds (see photographs 2a and b) and Sir Arnold Lunn saw a large iron bridge section thrown 150 feet into the air by avalanche blast. There are many instances too of men being plucked up and flung quite considerable distances.

In recent years attempts have been made to measure the impact pressures of avalanches by erecting gauges in known avalanche paths. The anticipated pressures were well short of reality and many gauges have been ripped out of their concrete foundations. But, in the relatively short time that successful results have been obtained, an impact pressure of no less than 22,000 pounds per square foot has been recorded. This was in the notorious avalanche gully of the Val Buera near Zuoz, Switzerland, in 1961.

To put this destructive power into perspective, it is interesting to record the damage caused by a much smaller avalanche whose impact pressure, according to Dr. Voellmy, a Swiss expert in avalanche destructive power, did not exceed 485 pounds per square foot.

In the stormy night of January 11th, 1954, a train was standing before the station at Dalaas in Austria when the avalanche struck. It lifted the 120-ton locomotive off the rails and slammed it against the station; it hurled the passenger carriages about, tossing one down a slope, and it demolished several buildings. Ten people lost their lives as a result of this small avalanche, small because it was 45 times less powerful than the one measured in the Val Buera.

Wherever snow lies on slopes the latent threat of avalanches exists, and even England has had an avalanche accident. It is commemorated now by the Snowdrop Inn in Lewes, Sussex, for on this site a number of cottages called Boulder Row were swept away by an avalanche on December 27th, 1836. Eight of the 15 people buried lost their lives.

An account in the *Sussex Weekly Advertiser* at the time read: 'A gentleman who witnessed the fall described it as a scene of the most awful grandeur. The masses appeared to him to strike the houses first at the base, heaving them upwards and then breaking over them like a gigantic wave to dash them bodily on to the road, and when the mist of snow, which then enveloped the site, cleared off, not a vestige of habitation could be seen—there was nothing but an enormous mould of pure white.'

But it is in the Alps and the Himalayas, in the Rockies and the Caucasus, the Andes, the Pyrenees and the many other mountainous areas of the world covered by snow in winter that avalanches exert their main influence. They claim human lives as far apart as Norway and New Zealand, Persia and Japan. The largest avalanches doubtless take place in the Himalayas, where climbing expeditions have reported that they reach stupendous proportions and are unbelievably vicious in their frequency. But in terms of human suffering, avalanches have afflicted the inhabitants of the Alps of Europe to a greater extent than anyone else. This is simply because nowhere else in the world is a mountain range with winter snow so densely populated. In the Alps there is a rich harvest of life and the White Death has reaped, and still reaps, in plenty.

* * *

The first mention of avalanches was by Strabo, a wealthy man of Greek upbringing who lived from 64 to 36 B.C. He devoted his life to travel and wrote a massive work called *Strabo's Geography*.

When writing of the Alpine passes he describes the horrors 'of falling into chasms abysmal' if one makes a slight misstep, of the 'dizziness which comes to all, including the beasts of burden, who travel the passes on foot' and ends up by saying: 'Accordingly these places are beyond remedy; and so are the layers of ice that slide down from above—enormous layers capable of intercepting a whole caravan and of thrusting them all together into the chasms that yawn below. For there are numerous layers resting one upon another, because there are congelations upon congelations of snow that have become ice-like; and the congelations that are on the surface are from time to time easily released from those beneath before they are completely dissolved in the rays of the sun.'

Evidence suggests that avalanches were among the troubles which beset Hannibal's army on its epic crossing of the Alps in 218 B.C. The two main accounts of the crossing, those of Polybius and Livy, do not mention avalanches specifically but they do describe circumstances which make them a strong possibility. Which pass Hannibal crossed remains one of the great unsolved problems of history but it is known that he set out with 38,000 soldiers, 8,000 horsemen and 37 elephants. On the crossing he lost no less than 18,000 men, 2,000 horses and several elephants.

On the climb to the summit of the pass they were harried and attacked constantly by local tribesmen and finally reached the top at 'the setting of the Pleiades' or late October. They were exhausted, but Hannibal cheered his men by pointing out the plains of Italy spread below. After camping for two days, they began the descent.

This descent proved even more disastrous than the climb. They were no longer under attack, but of the total losses sustained more than half were due to the natural difficulties encountered while the army tried to make its way down from the pass. There was fresh snow on a crust of old snow, which the animals kept breaking through and sticking in, and a landslide had carried away 400 yards of path; the new snow made a detour difficult. Men and beasts in their thousands slid scrabbling and screaming into the abyss.

The snow conditions described can easily lead to avalanches and the poet Silius Italicus (A.D. 25–101) certainly describes them in his

version of the crossing in the epic poem *Punici*, a somewhat dramatized and perhaps not over-reliable account of the Punic Wars.
The relevant verse of the poem runs, in free translation:

'There where the path is intercepted by the glistening slope, he
(Hannibal) pierces the resistant ice with his lance. Detached snow
drags the men into the abyss and snow falling rapidly from the high
summits engulfs the living squadrons.'

Another point is that a great many scholars have persuasive
arguments in favour of the Col de la Traversette as the pass used by
Hannibal. It is therefore interesting that a document of 1475 states
that the tunnel pierced near the top of the pass in 1470 was designed
to protect travellers from avalanches on the final slopes. Overall
then, it would seem that avalanches were almost certainly among the
hazards which caused such terrible losses to that bold and brilliant
general.

By the 12th century, pilgrims on their way to and from Rome
began to report on the horrors of the Alpine passes in winter. In
December 1128, Rudolf, the Abbot of St. Trond near Liège,
crossed the Great St. Bernard, and he left a most interesting
account which is preserved in the records of the monastery.

With his party he reached Restupolis (Etroubles) above Aosta
with 'difficulty that was next door to death'; and they were then held
up by snowdrifts until a way through was shown to them, and they
reached Saint Rhémy 'on the Mount of Jove itself, after a distance of
two German miles. Here,' he wrote, 'as though fixed in the jaws of
death we remained in peril of death by night and by day.

'The small village was overcrowded by the throng of pilgrims.
From the lofty and rugged heights above it fell often huge masses of
snow, carrying away everything they encountered, so that when
some parties of guests had found their places, and others were still
waiting near the houses, these masses swept the latter away and
suffocated some whilst crippling others of those in the buildings.
In such a continual state of death we spent several days in the
village.'

During these days the *marones*, as the local guides were called,
refused to lead the pilgrims over the Pass; but finally the payment
offered became large enough to tempt them and they agreed to open
the track. They wrapped themselves and their heads in felt, put
rough mittens on their hands and pulled on high spiked boots

while the travellers went into the church to pray. Rudolf of St. Trond continues:

'When these fervent devotions were taking place in church a most sorrowful lament sounded through the village, for, as the marones were advancing out of the village in one another's steps, an enormous mass of snow like a mountain slipped from the rocks and carried them away, as it seemed to the depths of Hell. Those who had been aware of the mysterious disaster had made a hasty and furious dash to the murderous spot and having dug out the marones were carrying back some of them quite lifeless, and others half dead upon poles, and dragging others with broken limbs.'

It is hardly surprising that when the pilgrims came out of church and saw the tragic procession they hesitated hardly a moment before fleeing back to Restupolis in terror.

Another report was written by John de Bremble, a monk from Canterbury who crossed the Great St. Bernard in February 1188. He did not mention avalanches, but what he did say now seems so amusing that I cannot resist quoting from the letter he wrote to his sub-prior, Geoffrey:

'Pardon me for not writing. I have been on the Mount of Jove; on the one hand looking up to the heaven of the mountains, on the other shuddering at the hell of the valleys, feeling myself so much nearer heaven that I was more sure my prayers would be heard. "Lord," I said, "restore me to my brethren that I may tell them that they come not into this place of torment." Place of torment indeed, where the marble pavement of the stony ground is ice alone and you cannot set your foot safely; where, although it is so slippery that you cannot stand, the death (into which there is every facility for a fall) is certain death. I put my hand in my scrip that I might scratch out a syllable or two to your sincerity—lo, I found my ink bottle filled with a dry mass of ice; my fingers too refused to write; my beard was stiff with frost and my breath congealed into a long icicle. I could not write the news I wished.'

John de Bremble seems to have had a better excuse than most people for tardiness in dealing with correspondence.

*　　*　　*

By the Middle Ages, countless migrations of people had populated the mountain valleys to a surprising extent and, from this time on, life in those remote villages was ordered to a great degree by avalanche danger. From the 15th century onwards there are numerous records of terror and disaster in those idyllic valleys of the Alps.

Early documents from the Dauphiné area of the French Alps are full of complaints about avalanches and the catastrophes they caused. Near Oisans, in particular, people dared not leave their parishes for the six months or more that snow lay on the ground; and many were killed on their way to the cow-stalls or Mass on Sunday.

A translation of a typical Latin text of 1450 runs: 'In the hamlet of La Pouture d'Ornon there was such an abundance of avalanches that all the hamlet was destroyed with all the surrounding properties. Fourteen or fifteen people who lived in the hamlet were killed and now there is but one family which continues to live there with great difficulty. In addition, inhabitants state that last year, 1449, there was such an abundance of avalanches near the houses of the hamlet du Rivier that nearly all the dwellings were destroyed: they believe that if the avalanches had not brought with them the large amount of timber, which by the grace of God they did, the complete hamlet and all its people would have been annihilated.'

The Latin words used to describe avalanches in these early texts are interesting and help to trace the origin of our present word. They were usually called *labinae* or *lavanchiae*. *Lavanchiae* is probably of pre-Latin origin, perhaps Ligurean, and is of the same root as *lave* which is the flowing of mud or lava. Much later, confusion with the French *aval*, towards the valley or downwards, crept in and produced our present word *avalanche*, which we have taken from the French. It can of course be applied to any material falling down a slope, but by common usage the word alone has come to mean a snow avalanche, while the words stone or ice usually prefix it when these other materials are involved.

The other Latin word *labinae* comes from *labi*, the Latin for slide or slip. Later, the partial interchangeability of the letters *b*, *v* and *u* produced many local words in the Alps like *lauie*, *lavina*, *lauina* and finally the present German word for avalanche which is *lawine*, brought into general use by Schiller and Goethe.

After 1450, recorded avalanche disasters became ever more numerous. Villages were smashed, some of them repeatedly, and hundreds upon hundreds of lives were claimed by the inexorable snow.

The area around Disentis in central Switzerland figures prominently in the catalogue of disaster. In 1459 the Church of Saint Placidus just outside Disentis was destroyed completely. By then it had stood for no less than 655 years—the case *par excellence* of avalanches biding their time. The same avalanche destroyed a number of houses and killed 16 people; but ecclesiastical buildings in Disentis seem to have been the favourite target. In 1754 a large airborne-powder avalanche came down into the valley, and the air blast, apart from hurling a granite trough a 'quarter of a league' ($\frac{3}{4}$ of a mile) also blew the cupola off the convent tower, 400 yards distant from the avalanche path.

The nearby village of Trun suffered in 1808 when a blizzard lasting only three days deposited 15 feet of fresh snow in the mountains and nearly 10 feet in the village. An enormous avalanche from Kluka, on the east side of the Punteglias Valley above Trun, destroyed the chalets of Zeniu, swept up the opposite slope of the valley and devastated a large forest. It recoiled to the east slope and tore up some more woodland. It returned to the west, then back to the east where it pulverized six cowsheds. It went back to the west again, for the third time, burying a farm full of cattle; then it went east again, where part of the mass flowed over some low hills. However, enough remained to flow west again for a fourth and final onslaught in which the houses of Trun were buried to their rooftops. It is unusual for avalanches to recoil so many times and it could be that fresh avalanches were being released each time the masses crossed the valley. In any event, such stupendous destructive energy must have been terrifying.

By the 16th century the taking of medicinal baths was already popular; such places as Leukerbad in the Wallis (Valais) of Switzerland, St. Moritz and Pfäffers were visited by many of the wealthy of Europe. Yet, the little village of Leukerbad, set in a hollow at the head of a lovely valley, has had to pay dearly and repeatedly for its gift of curative waters; for it has been ravaged and scourged by avalanches for centuries. In most winters they seethed right to the outskirts of the village, occasionally claiming some victims and houses, but in 1518 the snow hurtled into the village itself and laid it waste. Sixty-one people were killed.

In 1718, the same thing happened again and an inhabitant of the time, Stephen Matter, left a good account of the disaster. In

December 1717 it snowed non-stop for ten days, a light, fine powder snow; then in the night of January 16th–17th it snowed again, and rained. At about 10 a.m. on Monday the 17th an avalanche swept into the village outskirts and buried three young men, Johan Roten, Stephen Meichtry and Matthias Andry. Only at dusk were they missed, however, and the whole village then hurried to the rescue with lamps and sounding rods (rods pushed down into the snow to feel for bodies). They could find nothing and returned tearful and grief-stricken to the village at about 7 p.m., but an even greater disaster awaited them there.

Just before 8 p.m., of the same evening, a monstrous powder avalanche struck the village. There was a single crash as if just one house had been hit; but in that instant houses near and far, weak and strong—'strong like the Sommer house whose firm walls had appeared as though they could withstand any avalanche'—were thrown down and swept away. Fifty-two people lost their lives and nine horses were killed, though there was luckily little livestock in the village at the time.

The St. Laurentius Chapel, all three baths, all the inns, over 50 houses and many stables were destroyed. Two houses east of the church were left standing, though ringed by avalanche debris, but apart from these a mere handful of houses above the church were left unscathed.

The church-bell summoned the survivors and with the pastor, Johann Plaschin, an attempt was made to find those buried but still alive. A few were saved but by morning, weeping with fatigue and sorrow, they doubted that any more survivors could be found. At midday, 12 people were dug out dead near the church. Joseph Brunner and his wife had been killed praying in the chapel while their four children were killed in their home. The house was so shattered that one child was found in a meadow some distance away, tucked up in bed as if by human hands. In a house above the chapel lived Noe Loretan, his daughter Christina and her husband Stephan Brunner with their three children. The family was found on the fourth day after the avalanche. Stephan Brunner and Christina were still alive but she died shortly after her rescue.

Stephen Roten, a healthy and strong young man, was fetching wine from the cellar in an inn when the avalanche struck. He was found alive after eight days, though snow had filled the cellar so that he had no food. He lived a further eight days but then died owing to the fearful frostbite of his feet and legs. It was hardly hoped that all the bodies would be found before spring, but help came from a wide area and after ten days everyone had been recovered except the little girl in her bed in the meadow. She was only found in spring. In all, 55 people died in Leukerbad in those few days.

By 1720, two years later, the stricken but valiant people of Leukerbad had rebuilt the baths, several inns and some houses only to have them destroyed when the mortar was barely set. And again in 1758, for the third time in 40 years an avalanche plunged into the village.

Not only did Leukerbad suffer in 1720; it was a year of avalanches of the intensity still dreaded today throughout the Alps, a year of the 'big snows'. At Obergesteln, near Gletsch in the Rhone Valley, the much-feared Galen avalanche came down with unprecedented violence and killed 88 people and 400 cattle, as well as destroying 120 buildings.

'God! What grief! 84 in one grave.' This was the original inscription on the headstone of the communal grave in which all but four of the victims were buried. Unfortunately, it has since been replaced by an inscription in verse which lacks the stark and moving simplicity of the original, and which conveys less well the shock in a small village when a disaster kills a major part of the population. In 1852 a small avalanche from the Galen slope attacked Obergesteln in a subtle way: it only destroyed the bakery, but the resultant fire razed the village. In 1915 the Galen avalanche came down yet again and destroyed a dozen houses.

But reverting to 1720: the children in the small village of Ftan in the Lower Engadine had all gathered in one house to sing when an avalanche came down and destroyed it; 32 young lives were lost.

At Rueras, in the Tavetsch Valley above Disentis, 100 people and 237 head of cattle were killed, and 60 houses were destroyed. After a later disaster it was almost decided to abandon the village—the only case to my knowledge in which the mountain people nearly gave up their stoic struggle for survival.

Still in this same winter, 40 people were killed near Brig, 7 in the Fieschertal, 23 on the Great St. Bernard, and 12 in Randa, near Visp.

In Graubünden (Grisons), the most easterly canton of Switzerland, avalanche disasters have been especially well documented. Among the earliest was one in 1440 at Davos when two houses near the lake were destroyed and 11 people killed, though a Martin Schlegel was dug out alive after 24-hours burial. In 1569, an avalanche down the same track killed seven and smashed through the ice of the lake. A large number of fish, killed by the concussion, were thrown out on to the land.

In 1598, over a hundred people died in avalanches in Graubünden, and in 1606 and 1609 the area around Davos was again terrorized. Hans Arduser wrote a chronicle of the area in which he enumerated the avalanche disasters between 1572 and 1614. In 1606, according to Arduser, it snowed for three weeks and by January 16th there were 12 feet of fresh snow. Just before midnight 'the mountains and valley trembled and shook'. Seventy buildings were destroyed and damaged at Davos-Frauenkirch, including the church, and the rescue work continued for three days and nights. Seventeen people died though five, among them a 14-year-old girl, were rescued alive after 36 hours.

Arduser also wrote that in 1609, on Ash Wednesday, March 13th at 10 a.m., the people of Davos-Dorf had just sat down to breakfast when a 'gruesome, grisly snow avalanche' came down and killed 26.

The Prättigau Valley of Graubünden, in which the famous resort of Klosters lies, has also been the scene of many days spent in mortal fear while a blizzard raged, of rumbling and crashing in the night, and of long days spent searching for loved ones in their winding sheets of snow. In 1689, at 8 a.m. on Saint Paul's Conversion day, an avalanche from Calmur killed 16, and a second at midday knocked down 155 buildings and caused the death of 57 people in the village of Saas. A poem was written to commemorate the disaster but, of course, the occurrence is now lost in the mists of time as far as the population of Saas is concerned. Anyone passing through the village, however, or skiing down into the Prättigau Valley from the Parsenn area, will see the steep, wide-open slopes above the village of Saas. If they know of the disaster in 1689, they will gaze at those slopes in fascination. And they will realize, with a pang of foreboding, that Saas can be annihilated again at any time an avalanche chooses. It is a sobering and spine-chilling realization,

even for those who do not live in Saas and have no intention of doing
so; yet there are thousands of villages throughout the Alps where
the same menace lurks.

*   *   *

The disasters mentioned here are, in the main, just a few of the
gravest taken from the almost endless catalogue of death and
devastation wrought by avalanches throughout the centuries in that
part of the Alps now covered by Switzerland. The Austrian,
German, French and Italian Alps have similar catalogues; as
isolated examples, 200 people died in Savoie, Nice and Aosta in
1755, and from the famous *Montafon Letter* it is known that more
than 300 people were buried by avalanches in the Montafon Valley
of Austria in 1689. Nor must it be forgotten that, winter after
winter, smaller avalanches, though still 'gruesome' and 'grisly'
ones, snatched victims here and there.

It should be recognized, however, that the mountain people were
not always blameless victims of an act of God when avalanches
struck them down: they often brought avalanches on to their heads
through their own thoughtlessness.

The woods above their houses afforded them natural protection
from all but the mightiest avalanches—yet the people plundered
and ravaged the timber outside their back doors as firewood and
building material. It is known that the Urserental, a valley leading
towards the Furka Pass from Andermatt, was once densely forested;
and yet by 1870 the 4-mile-long valley was denuded and such an
avalanche inferno that the inhabitants scarcely dared to remain.
Moreover, the Alps were the livestock-raising area of Europe
during the Middle Ages, and indiscriminate grazing of young tree
shoots was doubtless a factor in the deforestation, as well perhaps
as a certain amount of timber felling to provide pastures near the
villages.

However, apart from the factor of timber, houses were often
sited in suicidal places and rebuilt in the same place when destroyed.
But even the inhabitants of the safest houses had to go out to their
work in the forest or stalls, or to attend church, and they were then
exposed to danger. For this reason, avalanches have often had odd
side effects on the way of life of a village.

In the Kleinwalserthal of Austria, for example, the pastor of Mittelberg wrote to his deacon in December 1710 to tell him that the part of his congregation from the hamlet of Baad refused to come to church, because of the avalanches which menaced the path. The bishop agreed that Baad should have its own small church, and so it became ecclesiastically independent. But the avalanches had the last say because one tore the steeple and bells off the new church in 1789.

Another curious consequence of an avalanche occurred in the Maurienne area of the French Alps in 1579. A feud developed between the villages of Albiez and Villargondran when an avalanche had ripped out a forest in one parish and deposited the shattered remains in the other. The dispute as to whom the timber then belonged became so acrimonious that legal judgement was sought.

*　　*　　*

Avalanches have probably claimed an even heavier toll of human life from armies who ventured into the mountains in winter than they have from the civilian population. Although it is quite possible that the first in the long series of military avalanche accidents took place during the Retreat of the Ten Thousand, across the Armenian highlands in 401 B.C., or during Alexander the Great's crossing of the mountains of Persia, the Hindu Kush and Chawkpass, the first definite evidence is that mentioned earlier in connection with Hannibal. Thereafter, the first military avalanche accident of any magnitude occurred during the private war between the Dukes of Milan and Uri in 1478. The Zürchers, as allies, rushed to the assistance of the Milanese. During their crossing of the St. Gotthard Pass '60 soldiers were wretchedly devoured by an unexpected snow avalanche which rushed upon them'.

During the Swabian War, in 1499, two more accidents took place. As a result of the Swiss Federation's alliance with France, many mercenaries came into Switzerland and crossed the Great St. Bernard Pass on their way to attack the Milanese in the cause of Louis XI. They were surprised by a large avalanche and 100 men died. At about the same time, Kaiser Maximillian was climbing the Ofen Pass towards Zernez in the Engadine with 10,000 soldiers when 400 were suddenly buried in snow. The avalanche must however have been very shallow and slow-moving for none was killed, though several were injured. There was apparently much laughter as they extricated themselves—most probably the laughter of relief.

In 1799 and 1800, the area in and around the Alps became a cockpit as the French Revolutionary Armies waged war against their enemies. The Russian general Suvarow fought a fantastic campaign through central Switzerland in the autumn of 1799. By then he was 69 years old and had been called out of retirement to lead the Cossacks; but he was as massive and erect a figure as ever. There was a pitched battle with the French on the St. Gotthard Pass and, when Suvarow saw one of his lines fall back before an attack, he rolled off his horse in rage shouting: 'Bury me here! They are no longer my children!' His men rallied at once, stormed to the top of the Pass, routed the French down the Schöllenen Gorge and followed them to the Lake of Four Cantons, only to find insufficient boats left to carry them across.

Undaunted, Suvarow led his army across three passes, the Kinzig Kulm, Pragel and Panixer, a feat of mountain warfare to match any in history. But the Panixer Pass was almost his undoing for new snow was falling as he set out on October 5th. The losses of the army were appalling. Their boots were already worn out, their clothes were in tatters, and they were exhausted by their previous exertions. Though the leaders of the army reached Panix on October 6th, it was four days before the last of the survivors reached safety. As the blizzard raged and avalanches swept hundreds of men away, hundreds more just lay down in the snow and died. About 300 pack animals were lost, and so were the remnants of the artillery.

Neither did Napoleon's army escape unscathed when crossing the Great St. Bernard Pass on its way to the Battle of Marengo in May 1800, even though the danger from avalanches was known. In his instructions to the advance guard General Lannes said: 'No one is to cry or call out for fear of causing a fall of avalanches.' But it was Marmont's artillery that released an avalanche as they dragged their dismantled field pieces up the Pass in hollowed-out tree trunks. Quite a few men were buried under 50 feet of snow.

Then, in November of 1800, Napoleon ordered Maréchal Mac-Donald to take the Army of the Grisons over the Splügen Pass into

1a, 1b, 1c. Airborne-powder avalanche in three stages of development

2a. Part of the 250 acres of woodland destroyed by an airborne-powder avalanche at Vinadi, Lower Engadine, in February 1962 (see page 8.)

2b. Detail of the damage at Vinadi. Note how the air blast has stripped the branches and snapped the tops of the few trees left standing

Italy, and this Scottish immigrant's son who became a marshal in the French Revolutionary Army was taught the lessons of snow and avalanches the hard way. He wrote in his memoirs: 'I had more natural difficulties to surmount than enemies to conquer. Avalanches had swallowed whole squadrons.'

One of MacDonald's officers, Count Matthieu Dumas, wrote *Memoirs of his own Time* in which he described an avalanche which struck the first part of the army:

'An enormous avalanche loosing itself from the highest summit, rolling with a fearful noise and gliding with the rapidity of lightning, carried off 30 dragoons at the head of the column who, with their horses, were swept away by the torrent, dashed against the rocks and buried under the snow. I was not above 150 paces from the spot and thought for the moment that General Laboissière and his officers had likewise been swept away.'

The Army was forced back by the three-day blizzard and by numerous avalanches, and the guides refused to make a second attempt to cross the Pass until MacDonald himself led the column. He made the soldiers clear away the walls of snow in which, from the first attempted crossing, 'many men were entombed'. Then, by his own example, he made the men forget their fear and follow him through the storm while avalanches threatened from all sides.

There is also an anecdote about a drummer swept into the Cardinal Gorge on this march. He was not killed and he drummed for several days in the hope of attracting rescue. Potential rescuers were too busy looking after themselves, however, and he finally died of exhaustion.

But these and later incidents of history all pale before the incredible avalanche catastrophes which overtook the Austrian and Italian armies in the Tyrol during the First World War. A conservative estimate is that 40,000 troops died in avalanches between 1915 and 1918. Other estimates have been as high as 80,000, but the true facts are hard to establish. They were suppressed by censorship at the time but Matthias Zdarsky, the famous skiing pioneer who was training Austrian Alpine troops and became an avalanche expert, wrote: 'The mountains in winter were more dangerous than the Italians.'

On December 12th, 1916, a snowstorm began which continued for two weeks. Avalanches swept whole barracks away; 253 men were killed by one avalanche on the Marmolada. Then, avalanches were used as weapons, both sides releasing them on to the enemy below by firing a shell into a slope laden with snow. As a means of mass-murder this was far superior to artillery alone. Zdarsky states that 3,000 Austrian troops were killed in one 48-hour period and that the Italian losses were at least as heavy. The total of Austrian troops in the mountains at the time was only 80,000, so avalanches claimed a high proportion.

Even though the true facts and figures of the avalanche catastrophes in that tough Tyrolean campaign are hard to unearth, the following quotation from W. Schmidkunz's book *Kampf über die Gletschern* (Battle over the Glaciers) gives a good indication of the horrors which avalanches added to the struggle. It was written by one who took part.

'The White Death, thirsting for blood, claimed countless victims in the mountains. Whole barracks filled with happy men, dashing patrols and marching columns, were buried in the raging avalanches which followed the blizzards. Hundreds upon hundreds were the men gripped by the white strangler. Here and there some were quickly rescued, while others remained for a terror-filled day with both feet in the grave. But these were rare occasions. The snowy torrents are like the deep sea; they seldom return their victims alive. The bravest of the brave are covered by the heavy winding sheet of the avalanche. It is no glorious death at the hands of the enemy; I have seen the corpses.

'It is a pitiful way to die, a comfortless suffocation in an evil element, an ignominious extinction for the Fatherland.'

2

The Coming of the Tourist

In recent years, most of the Alpine villages which have been devastated by avalanches time and again through the centuries have lavished money on protective measures, a subject to be covered in a later chapter. But, in winters of exceptional snow-fall, disasters still occur as, for example, in the aforementioned winter of 1950/51 when nearly 300 people were killed in the Alps and as many more injured. In 1954, avalanches killed 160 people, mainly in Austria; in 1972/73 121 people were killed in Europe, plus about 1,000 Chinese troops in Tibet, according to the *Hindustan Times*. But the fact remains that nowadays 80% of all avalanche victims are tourists, usually skiers.

The boring of the Alpine tunnels in the 19th century made winter travel safer; the building of avalanche defences has made villages safer; but the development of Alpinism and the building of ski-lifts and cable-cars has taken hordes of people right into the lair of the enemy. The last 50 years have, in general terms, brought a shift in emphasis: today it is no longer among the travellers and in the villages that the plunging snow finds its principal prey—it is among those who go to the mountains for pleasure.

Climbing and tourism developed quite late in the Alps. The local people, apart from the odd chamois or rock-crystal hunter, never ventured further than the grassy slopes immediately above the valleys, for they were obsessed by all manner of superstitions about malevolent spirits inhabiting the high places and dangerous animals roaming the crags. It was believed, for example, that a ruined city lived in by the souls of the dead was perched on the summit of the Matterhorn. And Scheuchzer, the famous Alpine traveller, scientist and Fellow of the Royal Society, wrote in his *Itinera Alpina* of 1723 that although 'some dragons were fables' he held, nevertheless, 'from the accounts of Swiss dragons, and their comparison with those of other lands, that such animals really do exist'.

As an illustration, too, of the fanciful beliefs that kept the people in the valleys, the legend of Pontius Pilate is classic. The legend ran that after Pilate's suicide his body was thrown into the Tiber, but such a bout of storms and rain resulted that his body was promptly removed and dropped into the Rhone at Vienne. Storms again declared his displeasure. After a similar experience at Lausanne, on the Lake of Geneva, it was decided that he should be banished once and for all to the little lake on Mount Pilatus near Lucerne. The weather there immediately began to deteriorate in the established manner, but Pilate was quickly exorcized. He agreed to remain quiet for ever except on Good Fridays when, dressed in his scarlet judgement robes, he would sit on a throne in the centre of the lake. Anyone who set eyes on him would perish within the year. Nor was anyone to tease him by throwing stones into the lake, or great storms would again result.

The government of Lucerne took the legend so seriously that they expressly forbade anyone to go near the lake and six clerics who tried to climb Mount Pilatus in 1387 were very severely punished. Not until 1585 was the legend finally brought to ridicule by a Johann Müller of Lucerne: he hurled stones into the lake, and shouted taunts at Pontius Pilate, without there being any sequel other than continued fine weather.

So it was left to outsiders to climb the peaks. In 1492 Charles VIII of France ordered his chamberlain, de Beaupré, to climb the Mont Aiguille near Grenoble, erect three crosses on the summit and say a Mass—a feat which certainly only a royal command could have induced him to carry through. But, in the late 18th century, scientists, botanists and geologists began climbing in the current quest for knowledge of nature. In 1765, the de Luc brothers of Geneva attempted to climb the Buet with a barometer and thermometer to test the boiling point of water. They broke the thermometer and can hardly have been captivated by the joys of mountaineering, for five years were to pass before they repeated the climb, this time successfully.

The first man to climb for no other reason than his own pleasure was Father Placidus à Spescha, a Benedictine monk from Disentis. He made the first ascent of several peaks in Graubünden (Grisons) in the late 18th century and has attracted the title of 'the father of mountaineering'.

Most early mountaineering activity, however, was centred around Mont Blanc, which had been so named by Pierre Martell of Geneva in 1742, when he also stated that it was perhaps the highest peak in the Alps. Then de Saussure, a Genevese scholar, fell in love with Chamonix in 1760 after he had walked and scrambled in the area. He offered a reward for the first climb of Mont Blanc. This attracted the first serious attempt, in 1775, by four local men; but they failed and returned almost snow-blind. Not until June 1786 did Dr. Paccard and Jacques Balmat, a local chamois- and crystal-hunter, succeed in reaching the summit.

Eight further parties then climbed Mont Blanc before August 20th, 1820, when the first avalanche accident, indeed the first accident of any kind involving Alpine tourists, took place.

Dr. Hamel, Aulic councillor to Alexander the First, Emperor of Russia, decided to climb Mont Blanc with a Genevese optician called Selligue who hoped to try out a new barometer he had made. They were joined by two Oxford students of Brasenose College, Joseph Durnford, later to become a clergyman, and Gilbert Henderson. Together they engaged 12 guides, and during the first day they climbed to the Grands Mulets, where there is now a modern climbers' hut. A violent storm broke out in the night, which continued through most of the next day and forced them to stay where they were for a second night.

The following day dawned fine, and there are two stories about the decision to continue the climb. Durnford said in his account that the guides were eager to go on, as was the whole party with the exception of Selligue. He 'had decided that a married man had a sacred and imperious call to prudence where his own life seemed at all at stake; thus he had done enough for glory in spending two nights perched on a crag like an eagle and that it now became him, like a sensible man, to return to Geneva'.

On the other hand, Julian Devouassouds, one of the guides who survived the climb, told the Comte de Tilly the following year that the guides wanted to return to Chamonix when the storm finally cleared. They were forced, however, to give way before 'the stubborn insistence of Dr. Hamel who accused them, and the two Englishmen, of cowardice, so precluding any sensible consideration or discussion of the matter'. Devouassouds related too that Auguste Tairiez, one of the guides who was killed, had a presentiment of death and several times during the preceding night threw himself into the arms of the friend beside him, weeping and crying out: 'It's all over with me! I shall be killed up there.'

Whatever the truth, the party set out on a beautiful morning, leaving Selligue and two guides at the Grands Mulets. They stopped for breakfast on the Grand Plateau and continued at 9 a.m. for the final part of the climb.

Hamel wrote: 'We were full of hope and joy at seeing ourselves so near the end of our laborious journey. The glorious weather which prevailed, the awful stillness which reigned and the pure celestial air which we inhaled gave birth in our souls to feelings which are never experienced in those lower regions.'

The sun went in behind a cloud and Joseph Durnford, who had been second in the column, stopped to tuck his veil up under his straw hat. Three guides and Henderson came by him, and Henderson remarked that he thought there should be a guide between them in case of accident. But Durnford replied that there was no danger, and he made no attempt to regain his original position in the column. He in fact preferred the sixth position because it was easier walking where the track through the new snow was better stamped. His veil and his idleness were to save his life.

The party was carrying a pigeon in a cooking pot which served as a cage, and the bird was cooing contentedly. Dr. Hamel was looking down and counting his steps, breathless and lightheaded, when the snow suddenly gave under his feet, and the whole party was flung down the slope.

The avalanche was only about 70 yards wide, but it carried them over 100 yards. The first three in the column, Pierre Balmat, Pierre Cairriez and Auguste Tairiez, were swept into a crevasse and buried. Devouassouds and Joseph-Marie Couttet, who were fourth and fifth, were flung across the first crevasse and into a second, half filled with snow. Devouassouds came to the surface three times during the downward rush and saw Tairiez's black gaiters disappear into the first crevasse at the moment of Tairiez's fatal burial. Couttet and Devouassouds were able to escape from their crevasse with nothing worse than bruises.

The remainder of the party were only partly buried and did not take the incident seriously, until it was realized that three guides were missing. A chaotic, unsystematic search ensued, during which they jumped down on to the snow in the crevasse, prodded with their poles, and shouted down the holes they made. The only reply was a mournful echo from the Grand Plateau. The cold finally drove them down to Chamonix where a hotel-keeper had seen the accident.

The following morning the relatives of those killed were sent for. Old Balmat wept at the loss of his son and Durnford wrote: 'The sympathy which we could not help displaying in the grief of the surviving relatives wrung all the inhabitants' honest hearts.' The two Englishmen left all the money they could spare at the time to provide for the bereaved families and later opened a subscription in Geneva. Dr. Hamel, according to Devouassouds, gave nothing.

Forty-three years later a macabre selection of human fragments appeared in the Glacier des Bossons just above Chamonix. Among them were pieces of skull complete with hair, an arm, a foot severed below the calf and a large piece of a man's back, not to mention a frozen pigeon in a pot and Dr. Hamel's compass, which was still in perfect condition.

This accident occurred in what is called the Ancien Passage, a place usually by-passed by climbing parties after the discovery in 1827 of the Corridor, a longer but safer route. But, in 1866, Captain Henry Arkwright, aide-de-camp to the Lord Lieutenant of Ireland, his guide and two porters decided to save two hours by using the Ancien Passage. They too sprung the waiting trap of poised snow and ice. Captain Arkwright's body was not found despite a ten-day search; it too appeared in the Glacier des Bossons, this time after 31 years.

From the mid-19th century the 'Golden Age of Mountaineering' was in full bloom, and most of the prominent figures were British. Men like Tyndall, Tucket, Hudson, Whymper, Mummery, Dent and many others whose names have immortal status in Alpine circles, led in the discovery of the exhilaration and satisfaction to be found in the struggle with rock and ice amid the splendours of mountain scenery.

Travels in the Alps of Savoy, the famous book by Forbes, appeared in 1843, right at the opening of the 'Golden Age'. It is interesting, therefore, to see what Forbes then wrote about avalanches. He described them as the 'greatest and most resistless of catastrophes which can overtake the Alpine pedestrian', and went on to write of the 'thousands of humble travellers, of hardy peasants who have fallen prey to this unforeseen and appalling messenger'. He then stated that very few casualties had been occasioned by it (the appalling messenger) among amateur frequenters of the mountains. But it is significant that when Forbes' book was re-edited in 1900, at the close of the 'Golden Age', Coolidge added the footnote: 'Unfortunately many amateurs have since perished by reason of avalanches.'

Indeed, avalanches took more lives among the great climbers than did falls. This was probably because of the tendency to climb snow slopes rather than the safer rock ridges. Sir Leslie Stephen, for example, on his crossings of the Oberland passes, used to go up and down on snow rather than rock. And Tyndall wrote of an avalanche which almost killed his party when they could have been safely on rock.

In July 1864, Tyndall climbed the Piz Morteratsch with the big and ugly Pontresina guide, Jenni, another guide called Walther and two Englishmen, Hutchinson and Lee-Warner. On the way down Tyndall wanted to stay on rock but Jenni chose a snow gully.

It broke away in a small avalanche and they were all carried with it. Tyndall ineffectively tried to drive his baton into the underlying ice to stop them. Their velocity pitched them across a crevasse and Jenni, who was at the rear, jumped into the next crevasse in the hope of braking their headlong plunge. Even though he weighed 13 stone he was jerked smartly out and almost crushed by the rope.

The men at the front were being tumbled and buried by the snow while Jenni at the rear kept struggling to his feet, digging in his heels and yelling:

'Halt, Herr Jesus! Halt!'

They were being carried rapidly towards another group of evil looking crevasses when they slowed on a flatter spot, and Jenni's exertions stopped them. They were two or three seconds away from the first of the gaping crevasses.

Many climbers were less fortunate, men like William Penhall and Andreas Maurer who were killed by an avalanche on the Wetterhorn in 1882. Penhall was only 24, a medical student who was already numbered among the great climbers, and Andreas Maurer was a modest and quiet guide of great competence who had already climbed in the Himalayas. In the best traditions of self-sacrifice, Maurer once stripped to the waist to try and keep an English client warm when benighted in a storm on the Aiguille du Plan. Strangely, when Penhall and Maurer set out from the Bear Hotel in Grindelwald for their last climb, Maurer gave his pipe to Emile Boss, the hotelier, and said calmly:

'Take it. I shall probably never ask for it back.'

Some of the most famous guides of the epoch suffocated in avalanches after many years of mountaineering experience. Joseph Bennen, who was called the 'Garibaldi of guides' by Tyndall, was among them. Bennen came from Laax in the Upper Rhone valley and he was a somewhat strange and lonely man who lived with his mother and sisters after his wife's early death. He had a wide forehead, narrow chin and goatee beard. From Whymper's engraving he appears almost frail in comparison to the bear-like men who were the other famous guides of the time.

Tyndall and Whymper both thought highly of him, though they also thought him incautious and sometimes indecisive, but these views were not held by other clients. They built him into something of a superman in their descriptions, seeming to ignore the weaknesses he must have had—weaknesses which finally killed him and one of his party.

There is no doubting his courage, however, for he was the first guide to attempt the Matterhorn while it was still believed haunted, an attempt which made him famous. Vaughan Hawkins, who employed him on that occasion, described him as 'a perfect Nature's gentleman' and praises his boldness, prudence and cheerfulness.

Bennen had a favourite expression when faced by a difficulty which seemed to block further progress:

'*Es muss gehen!*' (It must go!) he would say, before tackling the problem with renewed determination. The origin of this motto was the story he loved to tell of a Tyrolean who complained to his priest in the confessional that an overriding passion for women struggled with the religion inside him.

'Son,' said the priest, 'to love women and get to Heaven, that doesn't go!'

'Father,' said the Tyrolean, 'it must go!'

By 1864 Bennen was among the most sought-after of guides. He was 45, at the height of his powers, and had made several first ascents when Phillip Gosset asked him to lead a party on a winter climb of the Haut de Cry, a very minor summit above the Rhone Valley. It is said that Bennen needed money as he planned to marry again, and although he had never climbed in winter he agreed to lead.

The party included three local guides, Bevard, Nance and Rebot, and Gosset's friend Boissonet. They set out on February 28th and climbed until only a snow field some 800 feet high separated them from the summit ridge. The slope was steep, and wider at the bottom than at the top; it was like a couloir on a large scale.

Bennen, who had been laughing at their struggle through the deep fresh snow, became anxious and asked the local men whether this slope ever avalanched. They said that it was perfectly safe. Bennen, with his great experience, should have insisted on turning back at that point, but they climbed up the north side of the couloir until they were 150 feet from the top. They then began to cut horizontally across the slope to reach the east ridge.

Bennen had been uneasy all the way up the slope, and when they began the traverse he said that he was afraid of starting an avalanche. The local men took his caution for cowardice and Bevard and Nance led until, three-quarters of the way across, they suddenly sank up to their waist in snow. They struggled on, cutting a deep furrow, until after a few steps they reached firmer snow again.

Gosset wrote: 'Bennen, undecided, had not moved but when he saw the snow hard again he crossed parallel to, but above the furrow made by the Ardon men. Strangely, the snow supported him. I tried his footsteps but sank up to the waist at the first one, so I went through the furrow with my elbows tucked in to prevent touching the sides. The furrow was 12 feet long and as the snow was good on the other side we had concluded that the snow was accidentally softer there than elsewhere. Bennen advanced: he had made but a few steps when we heard a deep cutting sound. The snow-field split in two some 14–15 feet above us. The cleft was at first quite narrow, not more than an inch broad. An awful silence ensued and then it was broken by Bennen's voice: '*Wir sind alle verloren*' (We are all lost). His words were slow and solemn and those who knew him felt what they really meant when spoken by such a man as Bennen. They were his last words.

'I drove my Alpenstock into the snow and brought the weight of my body to bear on it. I then waited. It was an awful moment of suspense. I turned towards Bennen to see whether he had done the same thing. To my astonishment I saw him turn round, face the valley and stretch out both arms.

'The ground on which we were standing began to move slowly. I soon sank up to my shoulders and began descending backwards. The speed of the avalanche increased rapidly and before long I was covered up with snow. I was suffocating when I suddenly came to the surface again. I was on a wave of the avalanche and saw it before me as I was carried down. It was the most awful sight I ever saw. The head of the avalanche was already at the spot where we had made our last halt. The head was preceded by a thick cloud of snow-dust; the rest of the avalanche was clear. Around me I heard the horrid hissing of the snow and far before me the thundering of the foremost part of the avalanche.

'To prevent myself sinking again I made use of my arms in much the same way as when swimming in a standing position. Then I saw the pieces of snow in front of me stop at some yards distance; then the snow straight before me stopped and I heard on a grand scale the same creaking sound that is made by a heavy cart passing over frozen snow in winter. I instantly threw up both arms to protect my head in case I should again be covered up. I had stopped but the snow behind me was still in motion; its pressure on my body was so strong that I thought I should be crushed to death. I was covered up by the snow coming from behind me. My first impulse was to try and uncover my head—but this I could not do for the avalanche had frozen by pressure the moment it stopped, and I was frozen in.'

Gosset was lucky in that his hands were above the surface and that Rebot, who had been able to extricate himself, soon came and freed his head. The rest of his body had later to be cut free with an ice-axe.

Gosset's friend, Boissonet, was buried with his feet out but was already dead when released. The rope which led to Bennen disappeared vertically into the snow. They could not move it and Gosset wrote:

'There was the grave of the bravest guide the Valais ever had, and ever will have. The cold had done its work on us; we could stand it no longer and began the descent.'

Bennen's body was dug out three days later from beneath 8 feet of snow. Seven months later his watch was found, and it worked perfectly when wound. Bennen's mother almost went out of her mind when told of her son's death, and only a collection of 3,000 Francs subscribed by previous clients saved her and her three daughters from abject poverty.

Ferdinand Imseng, another famous guide, deliberately exposed himself to avalanches but twice in his lifetime. On the first occasion the avalanches played around him but spared his life; on the second, an enormous avalanche annihilated him and his party with vindictive violence, as if to make amends for allowing his previous temerity to go unpunished.

Imseng was a hunter who lived in Macugnaga, the village below the towering Italian face of Monte Rosa, a 9,400-foot wall visible even from the Gulf of Genoa. The face is swept by avalanches and in particular, there is a couloir, a white gash down which avalanches rumble ceaselessly. An ascent of the face meant crossing this couloir.

Imseng had gazed at the mighty wall for years, and finally he became obsessed by a burning ambition to climb it, even though all the other guides, the greatest among them, had declared it too dangerous to warrant serious consideration. But Imseng studied the face and dreamed of glory until, in 1872, he persuaded the Pendlebury brothers and the Reverend C. Taylor to employ him for an attempt.

At 2 a.m. they set out, with even Imseng's jaunty confidence shaken by the thunder of an avalanche which had just hurtled down. But the climb went well and they crossed the couloir without incident. Luck was truly with them though, for an avalanche broke just below them on one snow slope; and near the summit the snow slid beneath their feet without avalanching properly.

The fame Imseng had dreamt about was his overnight: he had succeeded where nobody else had even dared to try. The impetuous, energetic, likeable Imseng, the gay bachelor with his feather in his hat, his bow tie and his bright blue jacket, the man 'who would have climbed until every vein in his body burst rather than have yielded to another', was the brightest star in the Alpine sky for several years. He made several other great climbs before August 8th, 1881, when he was drawn fatefully back to the scene of his first great triumph.

He was to lead Damiano Marinelli, a pioneer of the Italian Alpine Club, a wise, kindly man and a fervent patriot, who wished to climb the Macugnaga wall of Monte Rosa and so become the first Italian to do so. It had been climbed but once since Imseng's first assault nine years earlier, and then by an Austrian. The Marinelli party was completed by an Italian, Battista Pedranzini, as second guide, and by a porter with wood and blankets. The August day on which they set out was overwhelmingly beautiful.

As they climbed, Imseng became anxious about the crossing of the couloir. He pointed out his bivouac site of nine years earlier but insisted on climbing higher this time. At about 5 p.m. they crossed the couloir with their hearts in their mouths and reached rock again with great relief, even though it was rock which the larger avalanches swept from time to time.

The porter stopped to drink from a cascade slightly off the line of ascent which the others were on. Suddenly he heard the shout: 'Avalanche!' He looked up and was just in time to see the three men blasted off the mountainside by an enormous avalanche which had dropped from the semi-circle of peaks above.

The men were hurled like chaff; the air was full of gigantic snow tentacles and flying boulders. Three days later the barely recognizable bodies were found at the foot of the wall. The flamboyant Imseng had stretched his luck too far.

Early mountaineering literature is filled with warnings about avalanches, of accounts of avalanches that rushed by just in front of a climbing party, or obliterated their tracks just behind them. And finally the greatest of all the early guides—the 'high priest' among them—Alexander Burgener, fell to the choking clutches of an avalanche.

His death proves that even guides of great experience and judgement cannot always know which slopes are dangerous. Alexander Burgener had climbed with men like Mummery and Dent; he had led them up some of the most terrible climbs in the Alps. He had climbed in the Caucasus with the Hungarian Déchy, and again with Donkin and Dent, and he had also climbed in the Andes with Güssfeldt. He was a legend in his own time, a man whose daring and exuberance, whose agility and strength, took him up places others thought impossible.

He was stocky and well-knit with a luxuriant beard, and eyes set deep in a face which showed clearly his immense will-power and dignity. Yet he had humour and a weakness for the good things of life, especially the Bouvier which he and Mummery used to drink when about to climb a difficult pitch, or when one had just been completed. On the Grépon, one of their great first ascents together, they came to an apparently bottomless crack, and in Mummery's words:

'We had to hotch ourselves along with our knees against one side and our backs against the other. Burgener at this point exhibited the most painful anxiety and his '*Herr Gott, geben Sie Acht!*' (Good God, be careful!) had the very ring of tears in its entreaty. On my emergence into daylight his anxiety was explained. Was not the knapsack on my back, and were not sundry half bottles of Bouvier in the knapsack?'

Burgener's snow-craft was exceptional. He and Mummery often waited through a night in the rocks rather than risk a descent on sun-softened snow. Yet on July 8th, 1910, when Burgener was 66 but still immensely strong and active, he was leading a party to the Bergli Hut in the Bernese Oberland after a blizzard which had lasted several days. Outside the Eismeer railway station he gazed in silence at the glaciers for a moment and remarked that he was worried about avalanches. It was muggy and the *Föhn* (a warm southerly wind) was blowing. Snow was sliding off the station roof and thaw-water gurgling in the gutters when they set out.

The party was labouring up the final gully just below the hut. Christian Bohren from the Concordia Hut, who happened to be at the Bergli at the time, went out to meet them, while Kaufmann, the Bergli hutkeeper, was preparing a hot drink for the new arrivals.

At 6 p.m. Bohren and Burgener were exchanging the greetings of old friends who have not met for a while. They were 6 feet apart and almost outside the door of the hut when the avalanche broke beneath them. With a monstrous roar it carried away the whole party, and Christian Bohren too. Kaufmann dashed out with brandy bottle in hand; where nine men had stood a moment before there was now no one.

He rushed down the slope and found three bodies on the surface, all cruelly injured. One of the porters, Rudolf Inäbnit, had a leg almost ripped off; it was only held by some skin, which he wanted to cut with his knife.

Late that night, when the six other bodies had been found, a cortège lit by acetylene lamps wound its way sadly downwards. Inäbnit died on the way. Of the nine men swept away by the avalanche, which had broken off at a point where the snow was no less than 8 feet thick, only two survived. One of these survivors was Alexander Burgener (junior), but, to the loss of his father and brother Adolf beneath the raging snow, was added the loss of an eye.

It was an ignominious death for Alexander Burgener. As the *Alpine Journal* said at the time: 'If any party of mountaineers can be safe his ought to have been. One can only call it fate which in a perfectly easy spot, in one mad, surging rush, hurled the great guide and his companions to their doom.'

Fate or not, I prefer to regard the accident as an illustration of the essential unpredictability of avalanches, even for guides and Alpinists of vast experience. And since Burgener many other famous Alpinists have been killed by avalanches. The last, before going to press, was none other than Dougal Haston, a mountaineer famous for his exploits on some of the world's most daunting peaks. He released an avalanche, and lost his life in it, while skiing in the Bernese Oberland in January 1977.

*   *   *

On January 2nd, 1899, two men, Ehlert and Mönnichs, both well-known Alpinists of the epoch, were crossing the Süsten Pass near Meiringen in central Switzerland, when they released an avalanche which killed them both. The importance of this accident is that both men were on skis and skiing had been but recently introduced to the Alps. This avalanche on the Süsten Pass ushered in a new era of plenty for the White Death, an era which we are still in today; for skiers are under a greater threat from avalanches than are climbers. They need the snow slopes for their sport, while climbers can stay on rock; and skis are usually more likely to release an avalanche than are boots alone. In addition, of course, skiers frequent the mountains in winter when snow is more plentiful and avalanches more numerous.

Between 1899 and 1919, 86 skiers were killed in avalanches. This figure excludes Alpine troops, needless to say. Eighty-six may not seem many in 20 years but it must be remembered that skiing was in its infancy with few adherents in those years. And the period also includes four war years which were lean for the Alpine tourist industry.

Today it is a rare year in which 20–30 skiers are not claimed by avalanches in the Alps, and to protect skiers from avalanches becomes ever more difficult. Every resort is aware of its responsibilities and has an avalanche safety and rescue organization, usually

coupled with the rescue service for injured skiers. Yet, despite all
their efforts, skiers are killed in increasing numbers.

Hosts of lowlanders invade the mountains each winter: hundreds
of thousands strong they come. On Friday evenings, trains, buses and
cars bring happy hordes from all the major cities of the Continent.
They fly from London, New York, Chicago, even Sydney or
Capetown, for a few weeks in the glittering Alps. They are whisked
up into the realms of snow by the immense network of ski-lifts and
cable-cars, a network which is being expanded year by year.

These happy, but often foolhardy skiers from the lowlands
generally see snow only as a source of pleasure and delight. They
have little inkling of the menace it also holds. True, they realize
that a leg can be broken or an ankle sprained while skiing, but few
connect with snow the hideous possibility of being buried alive,
crushed and suffocated.

It is far from the purpose of this book to wave a finger of admoni-
tion, and it would be entirely wrong to conclude from what is to
follow that anyone who fixes skis to his or her feet is immediately in
mortal danger. An examination of the facts, however, shows that
almost all the skiers who die in avalanches today do so as a direct
result of flagrant disregard for warnings from the local avalanche
safety organization, either through ignorance of the possible consequences or through sheer perversity. In an action tantamount to
suicide they offer themselves to the snow as living sacrifices. This is a
strong statement but one which is based on ample evidence from
almost every ski-area in the Alps. As examples I shall cite some
instances that have occurred in the Parsenn area at Davos.

The safety and rescue organization for skiers in this area, the
aforementioned Parsenndienst, is one of the biggest and best of
its kind in the world. It was among the first to be formed and it
pioneered techniques for protecting skiers from avalanches. These
techniques have been copied in ski-resorts everywhere.

After a snow-fall patrolmen render ski-runs safe by controlling,
with explosives, avalanches which endanger them; but until such
time as this can be done certain runs are declared closed. Notices to
this effect are displayed at every ski-lift and cable-car station, and
at every mountain restaurant in the whole Parsenn area.

Time and time again these warnings are ignored with tragic
consequences. Many are the skiers who have died in this magnificent

. The Swiss Federal Institute for Snow and Avalanche Research at the Weiss-
ᵤuhjoch above Davos. The top storey—of wood—was added in 1964/5

. Microphotograph of a cup crystal (depth hoar)

6. Large crystals of surface hoar in a shady location and after a series of cloud-free nights

5. This engraving by Düringer from David Herrliberger's *Topographie der Schweiz* of 1773 illustrates the belief—widely held until about a century ago—that avalanches were enormous snowballs that gathered ever more snow as they rolled down the mountain

skiing area because they took warnings lightly or were blissfully ignorant, as they set off down a closed run, of the gruesome death to which they were exposing themselves.

Two examples of accidents caused by this sort of behaviour occurred in 1961. After a period of fine weather, a blizzard set in on January 31st, and by February 1st the Parsenndienst were forced to close certain runs. By the morning of the 3rd there was nearly 2 feet of fresh snow, and only a few carefully patrolled runs were still open.

A married couple from Vienna, Robert and Leopoldine Thalhammer aged 31 and 33 respectively, set out with a relative, J.G., from the top station of the Parsenn railway to ski down to Wolfgang in the valley. Before putting on their skis they passed two large notices saying in English, French and German: 'Avalanche Danger —Do not leave marked runs.'

They knew the run to Wolfgang well, and by 10.50 hours they were outside the Parsenn Hut, a restaurant about half-way down. They conferred for a moment and the hut-keeper watched them. Then he saw that they were about to branch off the proper run and ski down into a gully called Stutzalp. He hurried to them and warned them of the danger of avalanches. He advised them very strongly to stay on the run, which was marked by stakes driven into the snow and well tracked by earlier skiers.

But the trio did not heed the advice. They were in a hurry because Robert Thalhammer, who had had a car accident on the way to Davos, was due to visit a doctor that morning. The Stutzalp gully was the quickest way to the valley and so, when the hut-keeper had left them, that is the way they went.

The two men were skiing in front, but they both came to a sudden halt when their ski-tips dug into some deep snow which had drifted against a hump a few feet high. They laughed and extricated themselves. J.G. climbed on to the hump, while Robert Thalhammer brushed the snow off himself and waited for his wife, who was making her way towards him.

J.G. suddenly heard a sharp crack behind him, then a hissing noise. He turned. Where the young couple had stood a moment before there was now nothing but a fast-moving wave of snow. J.G. rushed to the avalanche debris as soon as it stopped, but there was no sign of the missing pair. With dread in his heart, he called for help.

About one and a half hours later the rescue team of the Parsenn-dienst finally uncovered the victims from under 6 feet of snow. They were beyond the help of artificial respiration or heart injections.

That very same evening, at 16.00 hours, a Parsenndienst patrol-man received a message that a solitary skier was setting out from the top station of the Gotschnagrat cable-car above Klosters, and that he was heading towards the Cassana High Route, closed for two days past. The patrolman snatched up his skis and chased the skier through the gathering dusk. He shouted desperately, but if the skier heard he paid no attention. The patrolman returned and raised the alarm.

He was joined by two colleagues, and together they set out on a reconnaissance of the avalanche slopes which the skier was bound to cross. The blizzard had begun again; fog swirled in and their lamps lit up the single track a bare 6 feet ahead of them. At great risk they followed the track, but after three avalanches had narrowly missed them they decided that the danger was too great and turned back. With the storm increasing in intensity they were fortunate to reach safety at 19.45.

The next morning it was announced that Pierre Imer, a 21-year-old Swiss, was missing and the rescue attempt began. His body was found buried 2 feet deep in one of the several avalanches which had scoured the slope he crossed. On the way to his almost certain death, Pierre Imer passed three general avalanche warning notices and also four signs informing him that the run of his choice was closed.

It would be agreeable to dismiss such disregard for the warnings of the safety organizations as rare occurrences, but it is unfortunately not possible to do so. During a particularly dangerous weekend when much new snow had fallen, and when the whole Parsenn area had seethed with avalanches for many days, six different closed runs were used, some of them on several occasions. And there were many instances of people straying right away from the marker stakes of those runs which were open. A party of 60 climbed across a certain slope on the Saturday morning; that evening an English student of 19 was killed on the same slope, and a large section of the track made by the climbing party was obliterated.

The strange obstinacy of skiers in skiing down closed runs would be understandable were no other runs available for their enjoyment;

but there are nearly always safe runs open. At Easter 1963, the Parsenndienst closed the Drostobel run above Klosters. In avalanche times this run is a deathtrap, for it consists of a large catchment area feeding into a single gully. An avalanche from anywhere in the upper part rushes into the gully, and anyone in the gully at the time would be lucky to survive.

The conditions which dictated the closure of the Drostobel at Easter were very similar to those which had prevailed only three weeks before: a sudden rise in temperature around midday which softened the snow and made it unstable. On that occasion the run had also been closed, but a number of skiers continued to use it. Seeing other people on the run, a ski-instructor with his pupil also set out down it, even though they had walked by several notices saying the run was closed. The pupil, a young Englishwoman, ran rather high into a slope, and she released an avalanche which killed her.

Therefore, at Easter, determined to prevent a repetition of such tragedy, two patrolmen were picketed at the start of the run to point out verbally that it was closed and to direct skiers down an alternative route. Skiers arrived and began to argue; they tapped their heads significantly with a forefinger and shot off down the run when the patrolmen's heads were turned. Only the summoning of uniformed police finally called a halt to the skiers' folly.

Other organizations have similar experiences to those of the Parsenndienst, though it must be admitted that some of the lesser bodies abuse the 'Avalanche Danger' placards by leaving them out too long, almost as an insurance policy against an accident. Crying wolf in this way is bound to damage skiers' respect for warnings in all resorts, even in those where the organization takes great pains to assess the danger accurately and to warn accordingly.

In fact, organizations that cry wolf are few and far between and it must be stressed that avalanches are a minor threat to those who care to cooperate with the safety authorities of a ski-resort. It is a tragedy that those skiers who do lose their lives each winter could usually have enjoyed more years of this exhilarating sport—had they been prepared to heed those trying so hard to help them.

Not only do heedless skiers endanger themselves; though they may escape unharmed when they ski down a closed run after a fresh snow-fall, they leave tracks which are like a magnet drawing others into danger. The next skier believes the run to be open when he sees the tracks. He may take a slightly different line down a slope, or he may fall, and the concussion will set the snow masses in motion. A moment later he is in their cold embrace and living his last terror-filled minutes. Those in part responsible for his pitiful death may be supping a hot drink in the valley and congratulating themselves on having once again proven that a safety organization had needlessly closed a run.

So far, of course, reference has only been made to avalanches in the downhill-skiing areas, areas controlled by safety organizations. Skiers who quit these areas and tour joyfully among the peaks and glaciers *ipso facto* run a higher risk from avalanches. This minority are usually protected to some extent by a fair knowledge of snow-craft, or if not they employ a guide. They also carry, or should carry, a bare minimum of safety and rescue equipment.

It is too much to expect those far greater numbers of skiers today, who confine themselves to downhill skiing, to have much knowledge of snow-craft, but *respect* for avalanches would help to reduce the number of skiers killed. On April 12th, 1964, a group of inter-national Olympic skiers was making a film at St. Moritz. They ignored several warning notices and one verbal warning about a certain area, and shortly afterwards several of them were buried in an avalanche. Bud Werner, the charming and brilliant American, and Barbi Henneberger, a young member of the German women's Olympic team, paid with their lives for the collective irresponsibility of the group. These people had skied since childhood and perhaps thought their knowledge greater than that of the safety organization. It may have been so, but it is never worth trying to prove it.

It is a sad and bitter truth that those happy groups in the cable-cars who poke fun at the 'Run Closed' signs, often see a dead skier dug from an avalanche before they realize the true nature of the risk they run. They often see the repulsive ravages wrought by an ava-lanche on a dead friend or loved one before they realize that (in the ski-pioneer Zdarsky's words): 'Snow is not a wolf in sheep's clothing—it is a tiger in lamb's clothing.'

3
Witchcraft and Research

Considering the devastation caused by avalanches in the Middle Ages and the number of lives they took, it is not surprising that superstitions and strange beliefs grew up around them. People thought avalanches omnipotent and incontestable. At best, therefore, they were believed to be acts of God as He worked His divine, albeit abstruse purpose for the world; but more usually they were thought to be diabolical weapons of the powers of darkness.

Doubtless, the arbitrary and whimsical behaviour of avalanches gave added credibility to these beliefs; for avalanches sometimes do the most curious things. Mention has already been made of the child found dead in a field, tucked up in bed as if by human hands. The wreckage of her home was several hundred yards away. There are numerous instances too of houses being smashed to rubble, while china-cupboards with contents undamaged, clocks that still run, and other breakables, have been set gently down hundreds of yards distant. The *Montafon Letter*, which relates that 300 people were buried in the Montafon Valley in 1689, also reports that while a priest was taking the sacrament to the dying he was buried by one avalanche and promptly unburied by a second.

In the Kalanka Valley of the Grisons in 1806, a whole forest was brought down. The flying tree-trunks shot over a village without doing any harm, but a single tree was pitched upright on the roof of the pastor's house as if growing there. The possible significance of this occurrence was the talking point of the village for months.

Avalanches may single out a house for destruction while others close by are untouched, as happened at Göschneralp in Switzerland during the winter of 1951. In January an avalanche came down from the south side of the valley and a pile of snow some 20 feet high was left between the village and the head of the valley. On February 13th a second large avalanche came down, this time from the head of the valley and at right angles to the first. It leapt across the pile of snow on a cushion of air, diving into three streams as it

39

went. One stream hit the first house and scattered the pieces widely. It then jumped the next few houses in its path, without damaging them, and blew in the ground floor windows of a house further on.

The middle stream blasted a cowshed over a circle 100 yards in diameter, killing nine cows and six sheep, while the third stream made off towards the pastor's house. The pastor, who was teaching the school-children at the time, had taken them upstairs to his kitchen because the downstairs schoolroom was too cold on that bleak morning. The snow-stream tore through the wall in one place only, through the empty schoolroom, and burst out of the other side. It took all the tables and chairs with it and slammed them against the church, fortunately protected by a snow-drift. The stream was a mere 10 feet wide but it lacked nothing in power.

In the past, it was easy to explain such occurrences by saying that evil spirits controlled avalanches. If a house was spared while others were destroyed it was because the inhabitants were favoured by the spirits. It was believed, too, that these spirits dislodged avalanches in vengeance or spite and then stood on them as they thundered down, steering them with a tree held like a rudder.

There is a legend that at Erstfeld, in the Canton of Uri, an old woman dressed in black, wearing a bonnet and carrying a basket over her arm, used to wander out from the woods, through the village and out towards Wyler. Nobody knew who she was or whence she came, but there were soon rumours being whispered around the village. When a man remarked that his livestock had strayed in the night his friends would say, with significant emphasis: 'I expect *she* untied them.' If someone fell ill they remembered that *she* had passed by the house.

Then one day an avalanche stormed down the Wyler Valley and, by all that is Holy, the old hag was sitting on the first wave of snow, quietly turning her spinning wheel! That was the last straw—the evil witch had overreached herself. Just as she was thanking Providence for having spared her life, four men grabbed her up out of the snow and carted her up into a clearing in the Wyler wood. There they flung her on a heap of firewood and summarily burnt her alive.

In this connection there was a famous witch trial at Avers in the Hinterrhein Valley in 1652 at which it was stated quite categorically that 'witches are the causes of avalanches'. In parts of Graubünden (Grisons) in this era, eggs, on which the sign of the cross had been made, were dug into the snow at the foot of slopes as protection against avalanches and the spirits controlling them.

If we think, too, of the panic and terror inspired by the overwhelming and capricious snow, it is understandable that avalanches should have been personified into a fiendish being, an ogre of malicious intent.

'*Was fliegt ohne Flügel, schlägt ohne Hand und sieht ohne Augen—das Lauitier!*' (What flies without wings, strikes without hand and sees without eyes—the avalanche-beast!)

Mountain folklore includes a number of such sayings, as well as quaint names for avalanches and other tokens of the people's fear and respect for a destructive power beyond their comprehension.

* * *

Early writers about avalanches were usually men who had been forced to travel the Alps in winter, men like the monk Felix Faber of Ulm who in 1483 described two southward pilgrimages he had made over the Alps. Both he and Josias Simmler, who wrote his famous book *De Aplebus Commentarius* in 1574, confined themselves to describing the *menace* of avalanches.

Not until the 18th century did writers attempt to describe avalanches themselves, and to classify them into different types. It was usually considered that avalanches were made up of giant snowballs, or even one big snowball, and all the earliest illustrations of avalanches portray them in this way (see photograph 5). Even Immanuel Kant, in his book *Physical Geography* in which he gave a section to avalanches, perpetuated the idea of a small snowball rolling down a slope and becoming progressively larger as it went.

Generally, however, the writers of the latter half of the 18th century agreed in the division of avalanches into two main types: *Staublawinen* (dust or powder avalanches) and *Grundlawinen* (ground avalanches). These in fact were the terms originally coined by the mountain people themselves.

The first type, the *Staublawinen*, was the type mentioned in the first chapter, the dry, fast-moving cloud of powder snow with its associated wind-blast. Indeed, in some parts of the Alps they were called 'wind avalanches', or sometimes 'cold avalanches' because a prerequisite for the snow to maintain its powdery nature is a low temperature.

The name *Grundlawinen* was applied to the enormous wet-snow avalanches which often occur in spring and carry rocks and soil with them. They do little damage, in general, because they flow slowly down well-known tracks and gullies. These avalanches were also known as 'warm avalanches' in certain areas because they usually occur in times of thaw.

It would be as well perhaps to state at this juncture that, when referring to temperature in connection with snow and avalanches, anything approaching or slightly above freezing point is warm. The term cold is reserved for temperatures well below freezing point.

Clergymen from avalanche-stricken villages were among the most prolific writers about avalanches in the 18th century, but none of them made much attempt to discover what caused the snow suddenly to slide. They knew only that there was an increase in danger during snow-storms and during the period following them. Then, in 1770, J. A. de Luc, one of the Genevese brothers who climbed the Buet as mentioned earlier, wrote a book about their scientific findings and included a chapter on avalanches. His explanation of the phenomenon was a valiant attempt but not, as one would expect, entirely accurate. Indeed the first accurate *description* of avalanches, leave alone their causes, did not appear until 1849 in a book about the Alps by Johann Georg Kohl, city librarian of Bremen.

The true foundations of avalanche research were not really laid until the second half of the 19th century when Johann Coaz began his invaluable work. Coaz was a Swiss inspector of forests, and he therefore had an interest in avalanches as they affect woodland. But he was also attracted by the fascination of avalanches, a factor far more important than his purely professional interest.

Since Coaz's time, many other men have been similarly fascinated, enthralled by the majesty, power and unpredictability of avalanches. Most of these men have been Alpinists and skiers with a deep passion for the mountains. Then, as a result of seeing avalanches at close quarters, they have fallen under their spell and have gone on to spend many years carrying out research and observations, often as amateurs and with little chance of material benefit.

They have become completely absorbed by the intricate nature of this spectacular phenomenon and by its mysterious causes. Why, for example, does a snow-fall of a given depth sometimes produce avalanches and sometimes not? Why, when there is snow every winter, does a slope which has not avalanched in hundreds of years suddenly do so? Why is it that maximum avalanche activity is reached a few hours after a snow-fall, lasts a few hours and then starts to diminish again? How does it happen that several people can cross a slope in safety, and then, suddenly, it avalanches and buries the last few of a party? How does an airborne-powder avalanche develop its incredibly destructive wind-blast? These are a few of the many questions which have attracted such men.

The work of unravelling the answers is often baffling, occasionally dangerous and always exciting. It cannot be dull: the elusive and malevolent character of avalanches guarantees that. All the men who have been captivated by them have worked with the energy and purpose exhibited by Coaz, and even the professional research workers of today are fired by the same boundless enthusiasm. Not for nothing has Coaz been called the 'father of avalanche research'.

Before becoming a forester, Johann Coaz assisted General Düfour, chief of the Swiss Federal Topographic Bureau, in carrying out the first accurate survey ever made in the Alps, and in the subsequent preparation of an atlas. This activity caused Coaz to travel a great deal, as did his forestry work later. Inevitably, he saw numerous avalanches and, as his interest in them grew, he began to keep meticulous records of snow-fall, temperatures and other factors that he thought might be of importance in their creation.

In 1872, when Coaz was forestry inspector for the canton of Graubünden, he wrote to other cantonal inspectors asking them to cooperate in a nation-wide avalanche survey, but shortly afterwards a Federal Forestry Inspectorate was set up which quickly took over the work which Coaz had initiated. By 1878, statistics were being compiled in all the Alpine Cantons as part of a plan for protective measures against avalanches, a plan which had just been set in motion.

Johann Coaz produced his book *Lauinen der Schwizeralpen* (Avalanches of the Swiss Alps) in 1881. It is a large and comprehensive work which will for ever remain a valuable source of reference. He followed this with a book on the damage caused in Switzerland by avalanches in the winter of 1887/88. Then, in 1910, he wrote a fascinating book on avalanche statistics and defences in the Swiss Alps, based, of course, on the information correlated over the previous years.

In this book, Coaz reported that there were 9,368 avalanche tracks in the Swiss Alps, and that the largest number of these were in the Rhine Valley complex, which had 2,320. He stressed that his survey only covered tracks of avalanches which affected mankind in one way or another, and excluded those high among the peaks.

Of the total, 2,958 were tracks of *Grundlawinen*, 932 of *Staublawinen*, 34 of glacier avalanches and 5,444 were of varied avalanches. Of these tracks 2,192 carried just one avalanche a year, 5,294 carried more than one avalanche a year, while 1,283 only carried an avalanche every few years. The remaining 599 tracks only carried an avalanche at intervals of many years. Spring had the most avalanches with 8,435, winter next with 6,744, and then autumn with 2,301. This makes a total of 17,480 actual avalanches per year.

I have made the point that avalanches are notoriously unwilling to follow any rules of behaviour, and so statistics—best regarded with reservation at all times—should be treated with even greater caution when applied to avalanches. Nevertheless, Coaz's work was invaluable and it is intriguing to speculate, with some 17,500 observed avalanches in the Swiss Alps alone each year, how many there are altogether in the Alps—let alone in the other mountainous areas of the world which lie under snow in winter.

The First World War campaign of the Austrians and Italians in the Tyrol produced, apart from the catastrophic losses described earlier, two men who were later to contribute much to the fund of avalanche knowledge.

Matthias Zdarsky, the skiing pioneer, gained his knowledge in the hardest possible way: he has the unsought-after distinction of having sustained more injuries in an avalanche, and yet survived, than anyone else. The accident occurred in February 1916 when Zdarsky was in charge of a rescue column trying to save 25 other soldiers already buried.

In a horrifying account, Zdarsky describes how he received no less than 80 different fractures and dislocations, including half a dozen of the spine. The snow of the avalanche smashed and kneaded him to a pulp; it squeezed him until he thought his eyeballs were bursting from his head. He was lucidly conscious of each part of his body as it was racked and twisted until a bone or joint gave way under the irresistible stress. He could feel his shattered ribs grating on his backbone, and he thought his entrails were being drawn from his body. He was wishing for a quicker journey into the hereafter, as he laconically remarked later, when the avalanche suddenly spat him to the surface. He survived, but as a contorted wreck of his previous self. With a will of iron he spent 11 agonized years rehabilitating himself to the point where he actually skied again.

Zdarsky wrote several practical booklets and articles on avalanches, as did the other expert produced during those years, Colonel Bilgeri, who was also an instructor in mountain warfare with the Austrian army.

Modern scientific research, however, was really initiated by Professor Paulcke of Innsbruck, who also began with practical advice to skiers and mountaineers before turning to scientific experiments. He culminated his work, in the 1930s, with some vital discoveries about the behaviour of snow crystals. At about the same time, Gerald Seligman was carrying out the observations which resulted in his book *Snow Structures and Ski Fields*, an erudite and still standard study in which he wrote a good deal about avalanches.

In fact, the 1930s were the time of great impetus in avalanche research. The Swiss Federal Government in Berne appointed an Avalanche Commission of 15 members in 1932. The Commission was made up from scientists, engineers and foresters and its task was to establish a study programme for snow, avalanches and avalanche defence measures. By 1933 preliminary work had begun at five places in the Swiss Alps.

In 1934, a man who was later to assume a role of great importance in avalanche research became involved in it almost by chance. He was Robert Haefeli, a civil engineer by training who in later life came to be numbered among the world's greatest avalanche experts. He was born in Lucerne and during his boyhood he developed a love of the mountains and of skiing. He had several brushes with avalanches but, at that stage, his interest in them did not develop beyond the casual.

In 1926 he was an engineer on a dam-building project in Spain, and he realized how little was known about determining the load-bearing capabilities of different soils and rocks. From this he developed a very keen interest in soil mechanics. He had become an acknowledged expert on the subject by 1934, and the Swiss Avalanche Commission therefore turned to him with a query: Did he think that the tests that were applied to soil, notably for sheer strength, could also be applied to snow?

Haefeli was working on an earth dam near Basle when this query reached him, and shortly afterwards, in February 1934, there was a snow-fall in the area. He promptly cleared all the soil from the apparatus in his site laboratory and spent two days experimenting with snow. As a result of these tests he wrote a 30-page report to the Avalanche Commission, who were so delighted that they asked him to go to Davos for a few weeks the following winter so that he could pursue the matter. In Davos progress was again encouraging; and for the next winter, that of 1935/36, the support of a famous crystallographer, Professor Niggli, was enlisted. One of his pupils, H. Bader, went to Davos with Haefeli, and later they were joined by E. Bucher, another engineer. Subsequently, the team was further strengthened by the addition of men like J. Neher and O. Eckel.

During the winters until 1939, these men were engaged in research for which there were few established precedents. They were, therefore, forced to conceive and develop many of the test methods and items of equipment that they needed. Their facilities were not good, but they were working with a spirit of discovery which drove them on. And those years were a source of great satisfaction to them all.

By the winter of 1936/37, the premises for research had progressed from the soil laboratory on the dam site at Basle, below the winter snow-line, to a wooden hut at an altitude of almost 9,000 feet on the Weissfluhjoch above Davos, passing by way of an igloo in Davos itself. The hut at the Weissfluhjoch was naturally refrigerated from outside, and it was not a comfortable place in which to work. Haefeli and Bader would wrap themselves in blankets and spend hours at a time absorbed in their exacting tasks. In the end the cold caused Haefeli to suffer a kidney complaint.

Nevertheless, Haefeli, who later became professor of soil mechanics, snow and avalanche defences at Zürich Polytechnic, remembered those years at Davos and at the Weissfluhjoch with evident pleasure. It was most interesting, and at times amusing, to hear him talk about them. He recalled that, at one stage, he and Bader were trying to find a way of measuring the hardness of snow. Bader favoured using a revolver and seeing how far in the bullets went, while Haefeli favoured a development of the ram penetrometer principle used in soil testing. (The ram penetrometer is described in chapter 4.) The discussion finally came to an end after they had wasted much time looking for revolver bullets in the snow. The ram penetrometer was developed and has found universal acceptance.

Bader, who later became director of the U.S. Snow, Ice and Permafrost Establishment, had plenty of original ideas. One morning, when he and Haefeli had arranged to go out for some field-tests, Bader turned up in his dinner jacket, claiming that it was the only clothing he possessed to which snow did not cling. Haefeli points out that today's ski-clothes are of smooth material, and he therefore credited Bader with having set an important trend.

Few outsiders took seriously the work that the team was doing—having little idea about the complexities of snow, they could not imagine what research there was to do. So, Haefeli and his men were often the butt of gentle mockery, mockery which turned to hilarious ridicule when they bought several dozen ping-pong balls for a certain experiment. (The experiment was completely successful.) And Haefeli once received a postcard addressed in terms which can only be politely, and semi-accurately translated, as: 'Snow-befouler Haefeli.' But, as Haefeli said, the main advantage of not being taken seriously was that they were left alone to get on with their work—and get on with it they certainly did.

In 1939, the team produced a monumental work called *Snow and its Metamorphism*. This book contained sections on snow crystallography, snow mechanics, variations in the snow cover as a function of aspects of terrain, practical methods and tests used in the study of snow, and, written by Chr. Thams of the Davos observatory, a section about density, temperature and radiation in the snow cover. The importance of this book, and of the work leading up to it from 1934, cannot be overstated. And, fortunately, the Swiss authorities were quick to recognize the tremendous promise shown by those early results.

It was also during this period that the study of avalanches had its small beginnings in the United States. In the winter of 1937/38, the first snow ranger was appointed at Alta, Utah, with the purpose of protecting skiers. Alta was an old mining town in which avalanches had claimed hundreds of lives at various times in the past.

With the outbreak of war in Europe, the team working at the Weissfluhjoch took over the avalanche training of the Swiss Army. The avalanche losses of the First World War, even though not suffered by the Swiss, were an object lesson which gave urgency to such a training programme. Even now the lesson has not been forgotten among Alpine armies; the Swiss have a large Avalanche Company made up of 200 men with specialized avalanche knowledge. The duty of this Company is to protect and ensure safe passage for the rest of the Swiss Army in time of need.

Doubtless, too, with thousands of troops in the mountains, the increased avalanche menace was a factor in the Swiss Government's decision to replace the wooden hut at the Weissfluhjoch with a permanent and more elaborate building. Therefore, in 1942, the Swiss Federal Institute for Snow and Avalanche Research came into being. And Haefeli, who returned to his job at the Zürich Polytechnic shortly after the Institute was built—leaving it under the direction of Edwin Bucher—remarked that the 300,000 Swiss Francs necessary for the new building were easier to raise than were the 500 Francs for the wooden hut of the winter of 1936/37.

Today, the Institute is the only establishment in the world where avalanche research is the main line of study. The rectangular stone building (see photograph 3) is built into a slope just beside the Parsennbahn Weissfluhjoch station; indeed it is connected to the station by a covered corridor. In the Institute, surrounded by snow from October to June, a team of 30 people is employed. Among them are engineers, physicists, meteorologists, foresters and skilled technicians.

It is an agreeable place in which to work, set as it is high above the valley of Davos and in the heart of one of the best skiing areas of the Alps. The establishment is small enough to have remained intimate and peaceful, a building in which men preoccupied and fascinated by their work can push steadily ahead.

The south-facing windows of the main floor give on to pine-panelled offices into which the brilliant Alpine sunshine streams. The rear part of the building is taken up by cold laboratories with every facility for carrying out research into snow crystal structure and the properties of ice. In the basement there is a very well equipped workshop with small lathes and other machine tools which the technicians need when making experimental apparatus. And there is a dark room for the use of the full-time photographer. The whole Institute has an air of quiet efficiency and, like most things Swiss, it is kept spotlessly clean and orderly.

The Institute is at present directed by Dr. Marcel de Quervain, a man small in stature, but possessed of great intensity and drive. He is an extremely able physicist and he inherited his interest in snow, ice and avalanches from his father, a meteorologist who, in 1909, led the second crossing of Greenland after Nansen. His quick and lucid mind has made him responsible for much valuable work in the field of snow crystallography. His papers are masterpieces of clarity. However, like most scientists with their own lines of study who also administer an organization, he is desperately busy and hard-worked.

The scientific research of the Institute is divided into four sections covering: I. Weather, Snow Cover and Avalanches; II. Snow Mechanics and Avalanche Defences; III. Snow Cover and Vegetation; IV. Physics of Snow and Ice. In addition, there is an Avalanche Service dealing with such day-to-day matters as the avalanche warning system and accident reports.

Section I, Weather, Snow Cover and Avalanches, is led by a meteorologist with the apt name of Dr. Föhn. His section studies climatology, with emphasis on matters relating to snow-fall and avalanches. Quite apart from the routine keeping of meteorological and snow data records, the section is working in the field of water-shed hydrology using isotopes to learn more about the water balance and snow melt in typical Alpine watersheds. Much of this work is of importance to the hydro-electric industry.

In conjunction with a French team from the Institut de Mécanique of the University of Grenoble, Dr. Föhn and other Institute staff are also working on statistical avalanche forecasting. More will be said about this in a later chapter. In addition, Dr. Föhn's section is involved, through NASA, in the use of satellite and manned orbital space stations to record Alpine snow-cover conditions. These observations help to build up a clearer picture of available natural resources of which snow, i.e. water, is one of the most important.

Section II, Snow Mechanics and Avalanche Defences, is run by Bruno Salm, a humorous balding man who spends long periods in the cold laboratories with his test rig for putting snow samples under triaxial stress. For one of the important lines of study of his section is to determine how snow changes when put under load, as it is in nature.

Perhaps the work of most immediate importance carried out by Salm's section, however, is that concerned with studying and developing avalanche defences on the basis of scientific knowledge of snow mechanics. As we shall see, the field of avalanche defence is now a complex science, and the very high costs involved make it imperative that the risk of error be reduced to a minimum.

Still in the field of avalanche control, Salm's section is also responsible for studying the effect of avalanches on objects they strike. Thus, it is Bruno Salm's team that organizes the afore-mentioned measurement of avalanche-impact pressures.

In fact, the Russians, in the years from 1936 to 1939, were the first to study such impact pressures, but not as part of a coordinated research programme. Many of the Russian measuring structures were ripped out bodily, and when the Swiss began their experiments in 1952 they were also astounded by the strength of structure necessary to withstand avalanches. Nearly all of their structures have been destroyed, some repeatedly, despite being rebuilt stronger each time.

The principle used for measuring avalanche-impact pressures, that of the Brinell test in reverse, is simple. A small disc of pure aluminium, of which the specification has been checked by the Federal Laboratory for Material Standards, is placed in an assembly so that it lies against three steel cones projecting from a back-plate. The usual arrangement is for two of these disc units to be mounted behind a rectangular plate oo·1 square metres in area, and this in turn is fixed to a structure built in a known avalanche path.

When an avalanche strikes the gauge the small cones are driven into the aluminium discs. By measuring the diameter of the im-pressions made in the aluminium the impact pressure, expressed in weight per unit of area, can be immediately calculated. Each little cone will measure up to 10 tons with a relatively small error.

Sometimes a slight variation of this method is used; rather than mount a pair of small aluminium disc units behind a rectangular plate, several larger disc units, each with a surface area of 0·2 square metres, are mounted on the structure in the avalanche path. The effect is, of course, identical except that the snow strikes the disc units directly rather than a plate in front of them.

Another quite different device was installed in 1959. It is a 10-square-metre panel with 15 separate pressure-measuring plates built into it so as to give readings over a much wider frontal area of the avalanche. Each pressure-measuring plate is backed by four little copper cylinders, called 'crushers'. After an impact the pressure is calculated from the amount that these copper cylinders have been crushed. It would have been well-nigh impossible to anchor such an enormous panel had not an ideal site been found at the base of an avalanche gully, nearly 1½ miles long, near Engi, in Canton Glaris; the panel is supported from behind by a granite outcrop nearly 20 feet high.

The early pressure-measuring apparatus using aluminium discs suffered from potential inaccuracy owing to the fact that the impact of successive waves of an avalanche might be registered cumulatively. To overcome this possibility, the apparatus was modified in the late 1960s so that, after the first impact, the device either locks to prevent further indentations being made, or the disc rotates so that any further impact will be measured elsewhere on it.

In conjunction with the pressure-measuring devices, most of the installations now have a speed-measuring device as well. These have two contact points, from 30 to 70 metres apart in the fall line of the slope, that activate a timing device by cable. The reliability of the device has not been exceptional to date; and some of the results, when the device has functioned, seem odd. For example, in 1969, an avalanche in the Val de Crusch (Lukmanier) was timed at 160 m.p.h. by one device, and at 42 m.p.h. by a second device some 400 yards further down the slope. The upper device was certainly in a steeper place, but such a large speed differential leads to the suspicion that the first device was actuated by the blast associated with the avalanche, which was mainly airborne and measured this rather than the downslope speed of the whole mass of snow. Thus it seems probable that the speed-measuring systems at present in use are more suitable for avalanches flowing along the ground than for the airborne-powder type.

There are now about half a dozen of these impact-measuring

installations scattered about Switzerland by Bruno Salm's team. It may be many years before a really large avalanche obliges by striking the gauges, so waiting for results is a long-term policy. But it is a sobering reflection that the avalanche in 1961 down the Val Buera, which left more than 22,000 pounds per square foot registered on the apparatus as mentioned in the first chapter, was smaller by far than many of the avalanches seen there in the years before the measuring station was installed.

On the other hand, avalanches confined to gullies like the Val Buera are known to exert higher impact pressures than those running down open slopes. A figure of 5,000 pounds per square foot would seem to be an average for a large avalanche on an open slope.

As well as these tests in the field, scaled-down avalanches are produced on a slide ramp outside the Institute. The aluminium track is 20 metres long and can be angled from 30° to 45°. A quantity of snow, usually 5 tons, is released from behind a shutter at the top of the track. Monitoring its movement, and its impact at the end of its course, with sophisticated electronic equipment gives insights into what to expect from real avalanches. Such information is vital for the construction of avalanche defences such as galleries to carry avalanches over roads, walls to divert avalanches away from habitations, and in fact any sort of defence scheme that depends on taming an avalanche already under way.

Section II is also working in the interesting area of the effect of explosions on the snow cover as a basis for improving the control of avalanches with explosives, of which more will be written later. And in addition to its research work, members of Bruno Salm's team are responsible for a continual monitoring of the effectiveness of existing avalanche defences throughout Switzerland and for advising during the planning and construction of new ones.

Section III, Snow Cover and Vegetation, is primarily concerned with problems of reafforestation. A dense forest is the best of all protections against avalanches, provided that there is no large open slope above on which an avalanche could gather sufficient momentum to tear out the trees. Avalanche-defence schemes are therefore always accompanied today by reafforestation. But even on slopes where the snow cover is supported by fences and prevented from avalanching, it may still creep downwards, especially in spring. This creep erodes the soil and uproots the newly planted trees, so presenting a problem in their protection.

The section at the Institute which tackles this problem is led by a forester, H. In der Gand. They experiment with a wide variety of different measures as they try to find the ideal way of protecting the saplings; they carry out tests with terraces, stakes driven into the ground behind each tree, little fences behind groups of trees, tripods set over the trees and so on. As with all forestry work, conclusive results can only be obtained after many years of observation.

In der Gand and his men have established several plantations in various areas, and each is divided into 100 plots of 10 metres square. No less than 50 different methods of protecting the saplings are being tried, two plots in each plantation being assigned to each method as an extra check.

Section IV, Physics of Snow and Ice, is run by Dr. Walter Good. He is a man of modest origins who by sheer ability and application has worked himself into an eminent position in the field of snow and avalanche research. Born in 1932, he is young for the responsibilities and achievements that he bears with great charm and modesty. He was trained originally as a horticulturist, and while working as a landscape gardener began to study again, eventually qualifying in physics and chemistry from the natural sciences department of the Federal Technical College in Zürich in 1962. In 1965, he obtained his doctorate on the basis of a thesis on molecular structure, no mean feat for an ex-landscape gardener.

At the Institute, his work and interest in structures, textures and related fields has had natural applications in the field of analysing thin-cut sections of snow and ice samples. He has done some important pioneering in what is termed 'numerical parameters to identify snow structure'. Put simply, and crudely, the process is as follows: polarized light is shone through an area of a thin-cut section covering some 100–200 cut grains. An automatic scanning microscope and a special device coupled to it—the whole called a tomograph—measure the light intensity at some 60,000 points over the surface of the sample and record the data on magnetic tape. This tape is then used as the primary information for a series of computer operations that describe the sample by 21 parameters. In effect, a computerized picture of the sample is drawn up which gives it a precise and objective description. Clearly, close cooperation between this programme of snow structure identification and the physical testing of snow samples by Bruno Salm's section could one day lead to some important new information about the behaviour of snow, and hence about avalanche formation.

Walter Good's knowledge and ability in the areas of mathematics, numerical methods and electronics have also led to his involvement in designing a remote snow-depth measuring device, in testing apparatus for finding avalanche victims, and in the aforementioned computerized statistical avalanche forecasting.

The Avalanche Service, which runs the avalanche warning system and investigates and reports on accidents, is headed by Melchior Schild, a cigar-smoking and somewhat terse man of wide practical experience whose hobby is the training of avalanche rescue dogs. He bases his avalanche warning bulletins—about which we will hear more later in the book—on information sent in from 52 measuring stations scattered through the Swiss Alps. In addition, neighbouring Alpine countries also feed their data to the Institute.

In recent years, the Institute has become increasingly international. Scientists from Germany, Norway, Britain, Japan, Canada, Czechoslovakia, the United States, Italy, France, India and New Zealand have worked there for varying periods. And Institute staff frequently go to work at other research centres overseas. The result is a number of joint research ventures with national and international institutes around the world.

It is a great tribute to the staff that, with the amount of academic research going on, they yet maintain a very practical approach to the problems. Instruction courses are run every second year for people from all over the Alps. They come to learn how and why avalanches form, how to safeguard themselves and others against them, and how to organize avalanche rescue operations. They see many practical demonstrations in the field at which most of the staff assist, whatever their particular sphere of work may happen to be.

After a snow-fall, too, the staff are often out filming avalanches and examining the snow at the point where the avalanche broke away. And if an avalanche accident occurs in the area they rush to reinforce the Parsenndienst rescue organization.

The Institute itself lies near the top of a series of avalanche slopes. An enormous one was once started by shovelling snow from the roof, and not so long ago a small avalanche burst open the back door. The staff like to say that it was offering itself for research.

Above all, the atmosphere of the establishment is one of dedication, perhaps best summed up by the notice in three languages displayed on the door. 'Dear visitors,' it reads, 'please come and see us between 10 and 11 on Thursday mornings, and for the rest of the week let us work!'

4
Snow

The treading of the realms of snow in the high mountains has enriched the lives of many thousands. The calm of an early morning high on a glacier, as the sun soars over the sentinel peaks and fires the landscape into all its radiant glory, is a balm for the soul. This glistening world is made of snow, the delight of children, of skiers and of those who find its beauty spread unexpectedly outside their windows.

Yet snow is also the stuff of terror, of those smothering masses which throttle and squeeze to death, which pulp the body and twist limbs awry; and the properties of snow are as varied as its effects on man. It is a material of great complexity and there is ceaseless activity within that deceptively peaceful blanket covering the earth. The crystals inside the snow cover are constantly undergoing change.

The British, Japanese and Swedes were the first to begin snow crystal research, but not very long ago as is shown by the still varied nomenclature for the different types of crystal. However, much has been discovered about snow in very recent times and it would be impossible and beyond the scope of this book to do more than outline the facts. Therefore, only those aspects of our knowledge of snow which have a direct bearing on avalanches and their formation will be elaborated.

THE FORMATION OF SNOW CRYSTALS IN THE ATMOSPHERE

Snow falling from the sky comes in thousands of different forms. An American called Wilson Bentley photographed 6,000 different kinds of snow crystal. Yet he believed that he had but scratched the surface and that many further types existed.

Bentley's mother showed him a snow crystal under a microscope when he was a child and kindled a flame of wonder and awe within him. By the time he was 20, in 1885, he had begun to use a camera to record the beauty he was constantly discovering under his microscope, and from then until his death in 1931 his enthusiasm for this exacting task never waned.

Bentley's life-work was a book of magnificent photographs in which are shown all the beauty and symmetry which Nature has invested in even so small an object as a snow crystal. A profusion of delicate forms are depicted; among them are stars, rods, prisms, plates, needles and rods capped with end plates. All the forms, however, have one factor in common: they stem from the basic hexagonal shape of the ice crystal.

Snow is formed in the atmosphere when air containing water vapour, water in its gaseous form, rises and is cooled. When air is cooled, its capacity to hold water vapour diminishes and the vapour condenses on to tiny particles always present in the atmosphere. Cloud or fog is formed. The particles which form the nuclei for the droplets usually consist of dust or salt, probably produced by the evaporation of sea spray. If the temperature happens to be above freezing point the droplets grow by further condensation, or by coalescence, until they become too large to remain suspended, and they then fall as rain.

If, however, the temperature is below freezing, some of the droplets will freeze into minute ice particles. These ice particles then grow as more water vapour attaches itself to them by a process called sublimation. This process, mention of which will be made frequently, *is the changing of a substance from its gaseous to its solid state, or vice versa, without passing through its normal liquid state*; in this case water vapour changing straight to ice. As the ice particles begin to grow in the cloud, supercooled water droplets will also be present, but their quantity will be reduced as the ice particles develop. Finally, the ice particles become too heavy to float and they are precipitated as snow crystals.

The basic form of snow crystal is a tiny hexagonal plate but as it falls through the moisture-laden air it continues to grow by further sublimation. The plates tend to develop more quickly on the horizontal than on the vertical plane and a spike starts to grow from each point of the hexagon until a star is formed. Branching plumes then grow from the spikes, at 60° to them, until beautiful structures like those shown in Fig. 1 have developed. The crystals can also grow on the vertical plane, in which case hexagonal prisms or rods are

Fig. 1. Dendritic snow crystals

formed; should the growth then become horizontal, end-plates will be added to these rods.

The variations are infinite, because so are the variations in the conditions which govern the development of snow crystals. The greater the concentration of water vapour in the air, and the longer the crystal is falling, the greater its development will be. Hence temperature also plays an important part for it influences the water vapour holding capacity of the air. Sublimation rate, too, is affected by temperature.

A snow crystal may pass through air strata of differing temperatures and varying humidity; it may pass through cloud; its fall may be prolonged by wind, and so on. The permutations are endless. In general, however, it has been found that snow crystals reaching the ground at high altitude, after a fall of short duration and where temperatures are low, are small and of the basic plate or rod types. At lower altitudes, where higher temperatures prevail and where the crystals have fallen further, they are larger and further developed into the star forms, which, incidentally, are called dentrites.

During their fall to earth, several crystals may stick together to form a snow-flake and, at high temperatures, it is common for enormous flakes to be made in this way. And those little pellets like soft hail, which the Germans call *Graupeln*, are formed when snow crystals fall for some time through fog or cloud. Supercooled water droplets freeze on to the crystal and round it into a pellet. This type of snow, and very large flakes, are seen in lowland Europe and in seaboard areas of North America quite frequently.

In one snow-storm, the type of crystals which reach the ground at a given place may well vary hour by hour with changes in any one of the factors controlling their development. New snow can vary from very dry 'wild snow', through normal dry fluffy snow, to wet snow which forms a ball when squeezed in the hand. 'Wild snow' is very light and only falls in very cold, calm conditions. It lies incredibly loosely on the ground, flies about when walked through, and flows off a shovel like water. Wet snow, on the other hand, falls at temperatures around freezing point. For the purpose of this book the expression new snow or fresh snow, unless further qualified, will be used to denote the usual dry, fluffy substance made up of dendrites, or little stars.

With so many variables in the formation of snow crystals, one can readily grasp Bentley's view that the 6,000 different types he discovered were but a scratch on the surface of the possibilities. After a few minutes spent looking at illustrations of snow crystals one can capture too some of the delight he must have experienced, delight which caused him to devote his life to the study of snow. His book of exquisite photographs, *Snow Crystals*, which he published in conjunction with his friend Dr. Humphreys in 1931, appeared just three weeks before his death.

THE METAMORPHISM OF SNOW

As soon as a snow crystal reaches the ground it is subjected to a change of environment. From having been a separate entity during its fall it suddenly becomes a minute part in the mass of the snow cover. At once, each crystal begins to undergo a series of changes in its nature which, if it is above the permanent snow-line and other conditions are favourable, will terminate eventually in the formation of glacier ice.

The crystalline and structural changes that occur in the snow cover are known as metamorphism, and we shall examine three distinct forms of the process: *destructive*, *constructive* and *melt*. Of these, only destructive metamorphism is invariably present in the evolution of the snow cover, the others needing certain conditions before they can take place. (In the past, the evolution of new snow to glacier ice was often called firnification, from the German *Firnschnee*, meaning last year's or old snow. Now, firnification has a narrower sense, usually excluding the first—destructive—stage of metamorphism.)

Destructive metamorphism. Because the dendrites which reach the ground are essentially unstable crystals, the fine points of the stars immediately begin to sublime. The resultant water vapour re-sublimes nearer the centre of the crystal so tending to deform the stars into rounded granules. These granules have a smaller surface area than the original stars and for this reason the new snow layer gradually settles. Most people have noticed how a new snow-fall of 1 foot is reduced to say 8–9 inches after a few days (see Figs. 2 and 3).

A new snow layer of fluffy dendrites contains over 90% air trapped between the crystals, and the specific gravity of fresh snow is usually about 0·05/0·06. Destructive metamorphism produces from this a layer of snow containing about 70% air and with a specific gravity of 0·2/0·3.

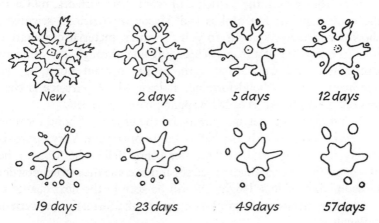

New 2 days 5 days 12 days

19 days 23 days 49 days 57 days

Fig. 2. The effect of destructive metamorphism on a dendritic crystal of new snow. This particular crystal was isolated and kept at a constant temperature of −5°C for the period of days shown

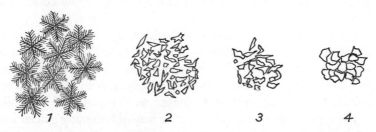

1 2 3 4

Fig. 3. Diagrammatic portrayal of the settling of new snow under the influence of destructive metamorphism

As the snow settles it also develops inner cohesion because the granules in contact with each other form bonds. This fusion of the particles is not strictly speaking part of the process of destructive metamorphism; it is however a function of the process because the rounding of the crystals allows them into closer contact and increases the contact area between them.

The process by which ice particles form bonds or necks was under discussion until quite recently. Some scientists, among them Dr. de Quervain and a Japanese called Nakaya, found that ice containing saline impurities may be covered by a thin, liquid-like layer. It is possible for the molecules in this layer to flow to the contact point, freeze and form the neck. Other scientists claimed that ice from within the particles, or from their surfaces, moves to the contact point; but Hobbs and Mason, two British scientists, showed conclusively about 10 years ago that material evaporating from the surface of the particles, diffusing through the ambient atmosphere and condensing again at the contact point, is the dominant agent in bond-forming, at least where pure ice is concerned. This process is called 'vapour-phase transport'.

Both destructive metamorphism and the process of bond forming take place more quickly the higher the temperature. Very approximately, it takes one to three days after a snow-fall in the Alps for the new snow to undergo these processes and become the settled powder snow in which skiers delight. As will be seen in the next chapters, destructive metamorphism plays a part in the formation of certain avalanches.

Constructive metamorphism. For this type of metamorphism to be understood something must be known about the snow cover as a whole.

Throughout winter, each snow-fall adds a layer to the snow cover and each layer will have different characteristics. The snow cover is therefore far from being a homogeneous mass. The layers will contain different types of crystal and different quantities of air. A layer of newly-fallen 'wild snow' may contain as much as 99% air while firn snow, old snow which is well on the way to becoming glacier ice, may contain only 20% air.

Especially important, though, is the temperature within the snow cover. The primary fact is that the ground under the snow is always at freezing point, or a mere fraction of a degree below. The upper layers of the snow cover, however, are much colder because they are affected by air temperature, which is usually many degrees below freezing in Alpine mid-winter. The practical effects of this are seen in the way a layer of snow protects plants and crops in the cold areas of the world. The prairies of Canada and the steppes of Russia would yield little were not the winter snow helping to keep the frost out of the ground.

This difference in temperature between the bottom snow layers and the top is progressive and is called the temperature gradient. It is directly responsible for constructive metamorphism in the following way: the snow crystals lying in the lower and warmer areas sublime continuously, and the resultant water vapour is carried upwards by the warmer air into the colder parts of the snow cover above. Here, the water vapour resublimes on to other crystals and causes them to grow. Thus the crystals higher in the snow cover are increasing their size at the expense of those lower down, and the number of crystals in a given volume of snow is decreasing.

In the first stage of growth by constructive metamorphism, single facets appear on the rounded granules, facets already showing the hexagonal characteristic of ice crystals. As the metamorphism continues, the angular crystals grow into curious structures like a cup with a pyramidal, stepped outer surface. The open end of this hexagonal cup is larger in cross-section than the closed.

These crystals can become quite large, up to half an inch in length, though a quarter of an inch is a more usual maximum. Occasionally the cup becomes filled by further sublimation. These strange crystals had been observed in polar regions in the 19th century but, in 1932, Professor Paulcke of Innsbruck was the first man to find that they existed in the Alps, and to reach conclusions as to their origin.

The all-important characteristic of these crystals in connection with avalanches is that they do not form bonds with each other owing to their angular shape. They constitute a non-cohesive mass in the snow cover. They are fragile and they run like loose pebbles; they form an unstable base for other snow layers.

The nomenclature of these crystals has, and does, vary. Paulcke called them *Tienfenreif*, which means 'depth hoar'. He chose this term because they are formed by sublimation of water vapour in the same way as air hoar. He also used the expression *Schwimmschnee*, meaning 'swim-snow', to characterize the free-flowing nature of the crystals. Many people in the Alps still call them this. In his book *Snow Structures and Ski Fields* of 1936, Gerald Seligman decided on 'depth hoar' as the best name. Today, on the Continent, they are often called 'beaker crystals', while in America 'cup crystals' is the name most used. I have decided to use the term 'cup crystals' as I think it the most graphically descriptive one.

Cup crystals then are an end product of constructive metamorphism. Their formation depends on air being able to circulate in the snow cover and so transfer the water vapour from the lower strata upwards; thus, anything like undergrowth in the snow cover which aids air circulation will intensify the process. And the steeper the temperature gradient (the temperature difference between the ground and the snow's upper surface), the greater will be the rate of cup crystal growth. They are usually found in the lower and middle strata of the snow cover, but in very cold weather, especially on north slopes and when there is little snow, almost the whole snow cover may be made up of cup crystals.

Melt metamorphism. This final stage of metamorphism concerns us less. It is the growth of ice granules by constant thawing and re-freezing and it produces firn snow, which skiers call 'spring snow' once the morning sun has softened it. Ultimately, melt metamorphism can produce glacier ice.

SURFACE HOAR

This is a brief outline of the metamorphism of snow, but the formation of one other type of crystal is very important to us, that of surface hoar. Like cup crystals, surface hoar is a product of sublimation. It is formed whenever the top surface of the snow becomes colder than the air above it, and when a certain concentration of water vapour is present. Such conditions occur on calm, cloudless nights when outgoing radiation from the snow can produce large temperature differences. This is especially true if clear weather continues for several days and nights. An early morning snow-surface temperature of $-20°C$ ($-4°F$) with an air temperature as high as $-5°C$ ($24°F$) is not uncommon. In these circumstances, moisture vapour in the air sublimes on to snow crystals on the surface and produces flat, leaf-like crystals of surface hoar.

Small crystals of surface hoar are often seen as those tiny glistening plates on the snow surface after a clear night. During the day the sun usually melts them, except of course on north slopes where they can become quite large. It is quite common to find surface hoar crystals up to 2 cm in size, though the largest ever recorded by the Institute at the Weissfluhjoch were 10 cm across. This was after an exceptionally long fine period during the winter of 1972/73. Should a snowfall bury a layer of surface hoar, it constitutes a menace because its crystals are very fragile and liable to break if overloaded. Subsequent snow layers therefore lie on a weak base, and many are the avalanches that have slid away on a stratum of surface hoar (see photograph 6).

PLASTICITY AND FIRMNESS

From the avalanche point of view there remain some generalities about snow which are important. Among these are the fundamentals that snow is a plastic material and that it also has firmness or strength in varying degrees.

That snow is plastic is obvious from the way it compresses when walked on or when squeezed in the hand, and the warmer it is the more plastic it becomes. Most people have noticed the squeaky noise that snow makes when walked on in cold weather and they may have noticed that this squeak does not occur in warm weather. The noise is in fact caused by the snow crystals breaking underfoot, which only happens when cold makes the snow brittle. The extra plasticity produced by warm weather allows the crystals to yield without breaking, hence there is no noise.

The firmness or strength of snow is a product of cohesion between the crystals, coupled with the resistance to breakage of the crystals themselves. The cohesion can be brought about in three different ways: firstly, by the existence of bonds between particles of settled snow; secondly, by the interlocking of the points and branches of dendritic crystals in new snow; and lastly, by capillary action in very wet snow. Each layer of snow in the snow cover may well have a different amount of firmness or strength, and avalanche research and forecasting is much concerned with this factor.

Cup crystals, for example, have very little strength, because they form no bonds and are fragile in themselves, while settled granular snow is very much stronger. The latter can also be strengthened still further by rolling; the compaction forces the granules into closer contact with each other and promotes the formation of firmer bonds. A practical application of this is the production of roads and runways from rolled snow.

An interesting aspect, however, is that to produce a lasting road or runway rolling must take place after every snow-fall from the first onwards. This is because rolling, by preventing air circulation in the snow cover, also restricts constructive metamorphism. If only the top snow layers are properly compacted, air circulation will allow constructive metamorphism to continue lower down, and the formation of loose, friable cup crystals will slowly undermine the road or runway until a vehicle breaks through.

Angular and cup crystals can only be strengthened by very high compacting pressures, and even then not to the same extent as granular snow.

In nature, snow layers are compacted and strengthened by the weight of other snow layers above them. This is particularly true of the first inches of a large snow-fall. Metamorphism will to some extent reduce the strength later—in fact a sort of struggle goes on between metamorphism and the factors trying to promote and preserve cohesion. But even so, a snow cover deposited in a few large falls will generally stabilize better than one deposited in many small falls spread over a long period.

THE EFFECT OF WIND

The carrying power of wind in the mountains is well illustrated in an account by Johann Coaz in which he reported seeing beech leaves being blown past him when the nearest beech tree was 16 km (10 miles) away, and several thousand feet lower.

During a blizzard, therefore, the tiny snow crystals are driven helter-skelter about the mountains by the shrieking wind. They are rubbed together and broken and few of them settle on the ground until they reach an area of calm. Here they form large drifted masses.

A curious fact is that these wind-deposited masses take on a hardness in varying degrees, from a slight toughness to being so hard that one can walk on the surface and leave no impression. The snow is said to be wind-packed.

There have been controversies as to why wind-blown snow hardens in this way, and it was originally thought that the pressure of the wind compacted it. Later it was held that the friction of the granules as they were blown along caused them partially to melt. This, of course, would produce moisture which would freeze after

7. The two basic types of avalanche: on the left is a very small slab avalanche and on the right a loose-snow avalanche (see page 78)

8. Small, wet, loose-snow avalanches initiated by increased thaw effect near rocks in spring

9. Fracture line of a typical slab avalanche on the lee-side of the cornice

10. A large *Grundlawine* in late spring. The avalanche has come down a gully behind the trees on the left of the photograph. Scale is given by the men digging a road through the debris

the granules had fallen to the ground, and so coment them to-gether.

In 1933, Gerald Seligman put forward his theory that wind-packing only occurs when the wind is charged with water vapour. The central point of his idea was that the water vapour condenses out on to the particles of snow, freezes and solidifies the mass. He argued his case very convincingly, and certainly no one has actually disproved it. However, it is now thought that mechanical damage to the crystals as they are being driven by the wind plays a very large part, for it allows the granules into closer contact so that they form bonds more easily. Indeed, snow which has passed through a rotary snow-plough hardens very rapidly, as does the snow of a powder avalanche, both cases in which the crystals have undergone mechanical damage.

Whatever its precise causes, wind-blown snow can form large wind-slabs up to several feet thick if a snow-fall is accompanied by strong winds. Even wind setting in some time after a snow-fall can transport snow and create slabs. These slabs of wind-packed snow can form very dangerous avalanches.

Wind also creates snow cornices. These often beautiful over-hanging masses have fascinated snow researchers, and much has been written about cornices and their formation. Apart from the ever-present possibility of a cornice collapsing and so causing an avalanche, our main interest is in a mass of wind-packed snow deposited below every cornice and called the scarp (see Fig. 4). Paulcke called it the *Gegenböschung*. Like all drifted deposits of snow in the lee of ridges it can avalanche very easily.

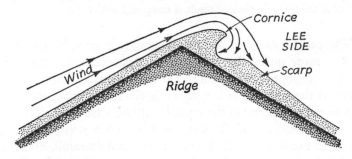

Fig. 4. A typical cornice and scarp formation. The scarp is a dangerous area of wind-packed snow

SNOW CREEP

Finally, the phenomenon of snow creep, the slow movement of snow down a slope, is important from the point of view of avalanches. Snow creeps because it has, to a certain extent, the properties of a fluid. Indeed, technically speaking, snow is classed as a 'viscoplastic' material, owing to the way it behaves in part like a viscous liquid and in part like a plastic solid. The viscosity of snow, like its plasticity, depends very largely on temperature: the warmer it is, the lower is its viscosity—that is to say, the more liquid-like it becomes, and the higher its creep rate will be.

The result of snow creep can be seen on the roof of a building a few days after a snow-fall. Originally the snow covers the roof in a layer of even thickness, but a few days later much of the snow has moved towards the eaves as shown in Fig. 5.

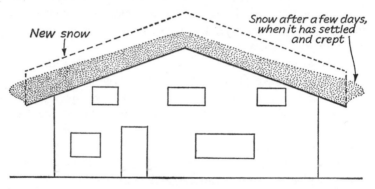

Fig. 5. The change in outline of a layer of new snow on a sloping roof a few days after the fall is an obvious result of creep and settlement

The process of creep is brought about, in part, by the settling of the snow under the influence of destructive metamorphism. Imagine a single snow crystal in the snow cover as settling is taking place. If the snow cover is lying on horizontal ground, the vertical downward movement of the crystal will of course be at right angles to the ground; but if the snow is lying on a slope, the vertical movement of settling will also take the crystal downhill. Equally, all the other crystals will be moving in the same way and in effect the snow will be creeping slowly downslope.

This creep movement caused by the settling of the snow crystals on a slope is increased by the force of gravity acting parallel to the slope. And movement brought about by this component of gravity can be very much greater if the snow cover is lying on smooth ground to which it is not anchored. The whole snow cover may then move *bodily* downslope, slipping along the ground; it is then said to be 'gliding'. (Note that in creeping, as opposed to gliding, the movement is caused by deformation *within* the snow cover and that the boundary layer along the ground remains stationary.) It is gliding of the whole snow cover as a body that does so much damage to vegetation and erodes the soil.

Creeping and gliding produce tensile stresses in the snow cover, and this is especially so on slopes whose vertical profiles are convex —slopes that are less steep at the top than they are lower down. On these slopes the snow on the steeper part tries to pull with it the snow on the flatter part above, so producing an area of tension within the snow cover (see Fig. 6). The same applies on slopes where the snow is anchored locally by rocks, undergrowth or some other feature of terrain. The effect of a *concave* slope is to create an area of compression within the snow cover because the snow on the flatter part supports the snow above it.

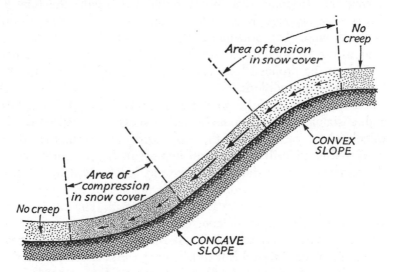

Fig. 6. The influence of a vertical slope profile on the forces in the snow cover caused by creep. The tension area is a potential fracture zone

THE OBSERVATION OF SNOW

A brief explanation has now been given of the processes inside the snow cover which gave rise to the earlier remark that 'there is ceaseless activity within that deceptively peaceful blanket'. Obviously, for the study of avalanches, a constant check on these processes is necessary throughout the winter. On the basis therefore of a system worked out by the Swiss Federal Institute for Snow and Avalanche Research, a series of careful observations is made, both there and at fifty-two other measuring stations, at fortnightly intervals throughout the winter. In Austria, Italy and France similar measurements are carried out by simular networks.

The first measurement taken is a penetrometer profile. This is a check of the firmness of the various layers within the snow cover. It has been found that the firmness gives a good guide to the strength of the snow, a factor which must be known for avalanche research. The instrument used, the penetrometer developed by Haefeli before the Second World War, consists of a pointed tube 1 metre long and 1 kilogram in weight, with a scale marked along it in centimetres (see Fig. 7). From the top of the tube a rod protrudes on which is mounted a sliding weight, also weighing 1 kilogram. By raising and dropping this sliding or driving weight the tube is forced stage by stage from the top surface of the snow right down to the ground. The principle is that of a pile-driver. Additional sections can be fitted to the tube for deeper snow, and the driving weight can be dropped from varying heights, or a heavier weight fitted for hard conditions.

The penetration of the tube at each stroke is noted. From this and a simple formula, a resistance, expressed in kilograms, is calculated for each layer in the snow cover. The results are then portrayed graphically and give an immediate visual impression of the firmness of the various layers (see Fig. 9, left of the upright).

Then a cut profile is made. Put simply this is the digging of a hole in the snow so that the strata can be examined. The cut is made along a smooth bottom of concrete prepared in the test area in summer. A neat vertical wall is cut and the total depth of snow noted. A coloured thread is then laid along the surface. This is also done after every snow-fall and the purpose of these threads is to demarcate the various strata as they occur. They also enable a check

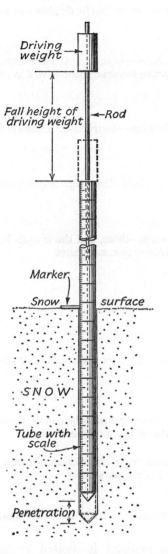

Fig. 7. Snow Penetrometer. The instrument used for measuring the firmness of the various strata

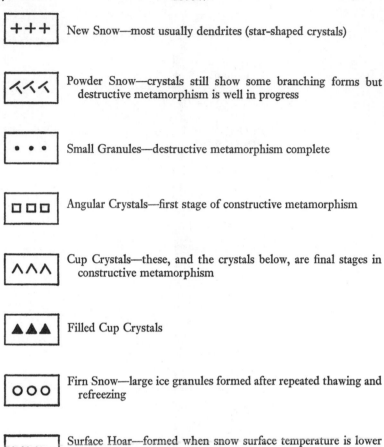

Fig. 8. The international snow classification symbols for the main types of crystal to be found in the snow cover

to be made of the settlement within the strata, as well as of the snow cover as a whole.

The temperature gradient is plotted by taking readings from thermometers inserted into the snow at 10-centimetre height intervals, and then the examination of the crystals in each layer begins. A sample is shaken on to a plate divided into millimetre squares and the crystals are examined under a magnifying glass. They are classified for size and type with great precision so that a

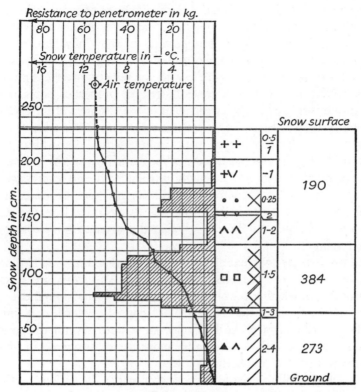

Fig. 9. A snow profile. In the left-hand part are shown the temperature gradient and the resistance to the penetrometer. On the right of the upright are the results of the cut profile showing the crystal types. The strokes and crosses next to the snow symbols denote the horizontal hardness of the strata. A single stroke means that the fist can be pushed into the layer, crossed lines—the fingers, double lines—a pencil, and crossed double lines (not shown here)—a knife. The size of the crystals in mm is given in the next column and on the extreme right are the snow density figures in kg/m³

comparison with earlier and subsequent profiles can be made in order to ascertain the metamorphism rate.

The specific gravity of each layer is measured by filling a tube of known cubic capacity with snow from the stratum under observation. The tube must be filled with care to avoid compressing the snow and so giving a false result. An aluminium plate is driven horizontally into the snow along a stratum-dividing line and the open-ended tube is pushed gently down from above, like a giant apple-corer, until it reaches the plate. The core of snow withdrawn is weighed, the test being done in duplicate for greater accuracy.

For avalanche study it is important to know how the snow is behaving from the point of view of tensile strength, which is similar to internal cohesion. The profile with the penetrometer has given a good indication as to the general strength properties but a more accurate check, which is independent of the hardness of the crystals themselves, is needed. To obtain this a core of snow is taken from a certain layer and centrifuged in a tube until it breaks and flies apart. From the revolutions per minute of the centrifuge at which the core broke down the tensile strength is calculated and expressed in grams per square centimetre. From the beginning of the winter until the end, the sample to be centrifuged is always taken from the same stratum so that it can be seen whether the strength is increasing or decreasing.

Two final observations are made, one for the horizontal hardness of the strata, and the other for moisture. The strata are graded according to whether the clenched fist can be pushed into them, the end of the fingers held flat, the blunt end of a pencil, or a knife. The visual moisture check divides snow into several categories, from the driest with no discernible moisture to the wettest in which there is more free water than snow.

A set of symbols exist for denoting on paper the various types of snow crystals and the results of the different observations. The main symbols for crystal type are shown in Fig. 8. There are others for expressing subtleties with which we do not need to be concerned. By using the symbols, a complete description of the snow cover at the time of the profile can be put on paper.

Figure 9 shows such a profile. It can be seen that the ground temperature is at freezing point, and it is also interesting to compare the penetrometer results with the type of crystal to be found in each layer. It can at once be seen that cup crystals have little resistance, hence strength, because of their aforementioned inability to form bonds, and because of the brittle nature of their structure.

The profile shown is the sort which could easily provoke avalanches. The layer of surface hoar is the weakest point and all the snow above it could slide away, helped by the cup crystals below acting as ball-bearings.

Profiles like this taken every two weeks are placed side by side to form a so-called time-profile, that is to say a series of profiles from which it is possible to follow the build up, settling, metamorphism and final melting of the snow cover from autumn to spring. To this can be added the daily weather readings of snow-fall, cloud cover, sunshine, wind, humidity and temperature, and a complete picture is built up.

It is only by such methods that an effective avalanche-warning system has been created in several countries. To predict when avalanches might strike, one must plumb the depths of the snow cover. Conditions on the surface alone are not enough, for the snow beneath is seething with activity; cup crystals have been built which are undermining the stability of the snow cover; a layer of fragile surface hoar has been snowed in; a recently stable layer has begun to loosen under constructive metamorphism. Only pains-taking observation of this fascinating and changeable material snow will give an insight into the mysterious ways of avalanches.

5

The Forms of Avalanche

Avalanches come in different forms and different sizes. As Joseph Wechsberg wrote in his dramatic book about the 1954 disasters at Blons in the Grosswalsertal of Austria: 'White Death has many faces.' In the past, there was a tendency for every expert who wrote about avalanches to propose a different classification scheme for them, and this led to confusion and argument. Today, work is still going on in scientific circles to refine avalanche classification, notably by a working group of the International Commission of Snow and Ice (ICSI).

Among non-specialists, avalanche description is usually very loose and still uses terms that originated long ago in mountain communities. The classic *Staublawine* (dust avalanche) and *Grundlawine* (ground avalanche) classification, which is as old as the settlement of the Alpine valleys, still persists among some mountain people. In fact, the first outsiders' attempts to classify avalanches in the 19th century were merely subdivisions of this traditional distinction. In 1889, for example, F. Ratzel, a German, distinguished two types of *Grundlawinen*—rolling and sliding—and in 1891, V. Pollak of Austria divided *Staublawinen* into 'pure', which were completely airborne in their movement, and 'common', which flowed in part along the ground.

In the 1920s classification became more ambitious. Some writers began by classifying from the kind of snow in the avalanche, others from the type of movement of the avalanche, while A. Allix of France produced a scheme which combined both and brought in several other criteria as well. Paulcke and Seligman later suggested schemes based on snow type.

There are several difficulties in avalanche classification. Among these is the fact that many commonly used terms, mainly in French, German and Italian, are deeply rooted in traditional avalanche parlance, are often untranslatable into other languages, and lack precision in their own. For example, Italian has two words for avalanche: *valanga* and *slavina* which come from different regional and etymo-

76

logical roots but are both now used nationwide. Ask the average Italian skier what the difference is between the two and you get many different answers, but most hint that a *valanga* is large and a *slavina* small, and that the former rolls while the latter slides. But specialists do not apply either of these criteria to avalanche classification.

In addition, classification is difficult because, as Dr. de Quervain points out, avalanches are physical *objects*, that can be described and photographed, but they are also *events* brought by influences of weather and snow-fall, the incident which starts the snow moving, and so on.

Among specialists today, the most widely used avalanche classification is the one that appears in Fig. 10. It was first proposed by Professor Haefeli and Dr. de Quervain in 1955. The first-ever international symposium of avalanche specialists, which met at Davos in 1965, discussed the classification at length, agreed on its practical usefulness, but recommended that more work go into it with special reference to its computerization possibilities and to classifying avalanches by their causes also. In late 1973, the aforementioned ICSI working group, under the chairmanship of de Quervain and made up of de Crécy (France), LaChapelle (U.S.A.), Lossev (U.S.S.R.) and Shoda (Japan), made proposals for two avalanche classifications: a morphological one to describe visible and measurable features, and a genetic one to describe the circumstances and conditions leading to a given avalanche of a certain morphological type.

The morphological classification proposed is a slight refinement of the one shown in Fig. 10 in that it subdivides slab avalanches into hard and soft slab and adds a few criteria such as the coarseness of the avalanche deposit and its contamination by stone, earth, trees, etc. It also includes an alphabetic and numerical code to allow an avalanche to be described by letters and figures.

The genetic classification is still not quite complete at the time of writing, but it sets out to cover fixed circumstances such as altitude, slope orientation and steepness, and ground cover; and such variable circumstances as snow-fall quantity, type of new snow, snow-fall intensity, wind, temperature, old snow conditions on the surface and lower in the cover, triggering mechanism, etc. The work is not finished owing to difficulties connected with classifying the triggering mechanism, which, as we shall see, is extremely complex.

AVALANCHE CLASSIFICATION SYSTEM

Avalanche Definition: Dislocation of the snow cover over distance greater than 50 metres.

CRITERION	ALTERNATIVE CHARACTERISTICS AND NOMENCLATURE	
1 TYPE OF BREAKAWAY	From Single Point 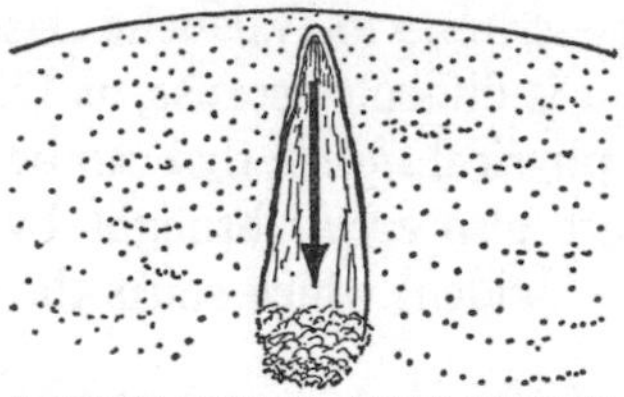 **LOOSE-SNOW AVALANCHE**	From Large Area Leaving Wall 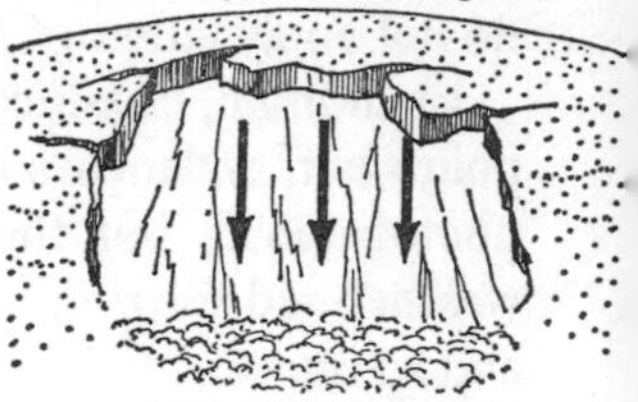 **SLAB AVALANCHE**
2 POSITION OF SLIDING SURFACE	Whole Snow Cover Involved 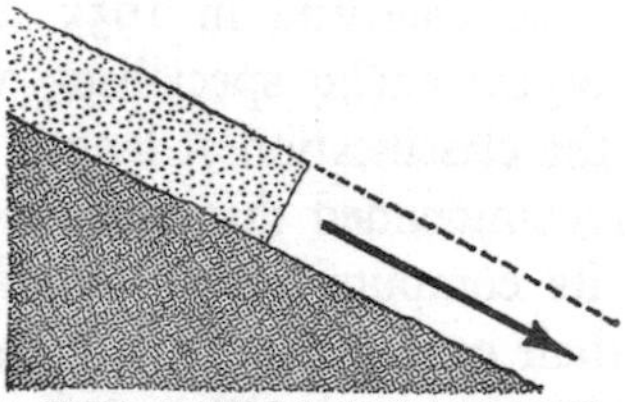 **FULL DEPTH AVALANCHE**	Some Top Strata only Involved 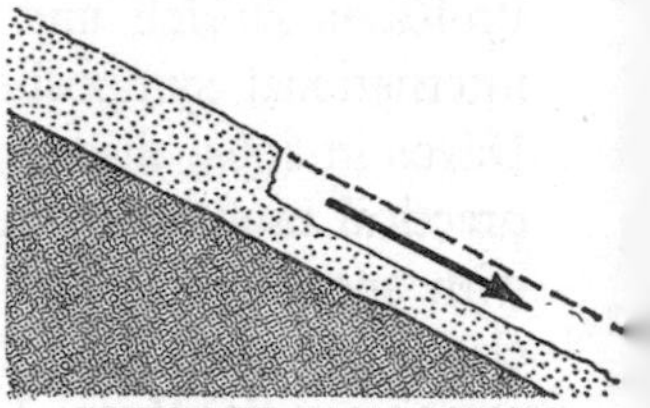 **SURFACE AVALANCHE**
3 HUMIDITY OF THE SNOW	Dry **DRY-SNOW AVALANCHE**	Wet **WET-SNOW AVALANCHE**
4 FORM OF THE TRACK IN CROSS SECTION	Open Slope 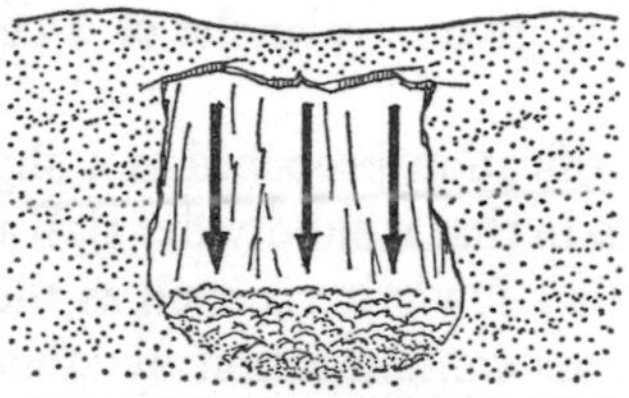 **UNCONFINED AVALANCHE**	In a Gully 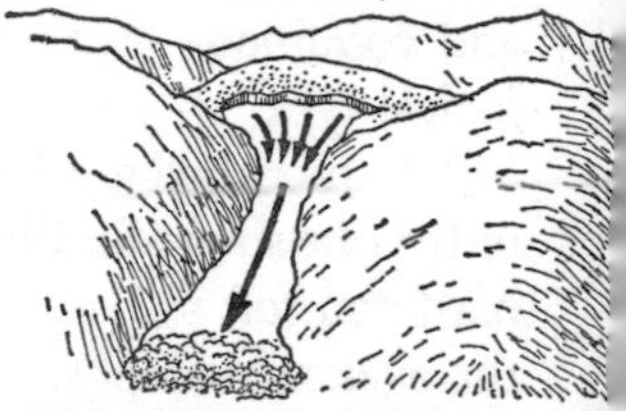 **CHANNELLED AVALANCHE**
5 FORM OF MOVEMENT	Through the Air 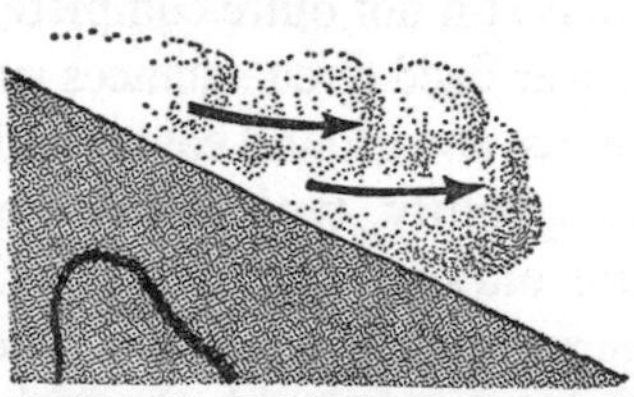 **AIRBORNE-POWDER AVALANCHE**	Along the Ground  **FLOWING AVALANCHE**

Fig. 10.

*After the system proposed by Prof. R. Haefeli and Dr. M. de Quervain
the Swiss Federal Snow and Avalanche Research Institute in 19*

However, despite all this avalanche classification activity in specialist circles, the Haefeli/de Quervain model of 1955, shown in Fig. 10, is more than adequate for our needs. It begins by defining an avalanche as a dislocation of the snow cover over a distance greater than 50 metres.

Criterion 1 is the form of break which started the avalanche and this leads to the broad division of all avalanches into two types: 'loose-snow avalanches' starting from a single point and setting ever more snow in motion as they go, and 'slab avalanches' which break off from a whole area all at once and leave a jagged fracture line.

Criterion 2 differentiates as to whether the whole snow cover right down to the ground has moved, in which case it is termed a 'full-depth avalanche', or whether just a few of the upper strata have moved, in which case it is termed a 'surface avalanche'.

Criterion 3 is whether snow forming the avalanche is dry or wet, producing the nomenclature 'dry snow' or 'wet snow avalanche', while criterion 4 is whether the track of the avalanche is 'unconfined', as it would be on an open slope, or 'channelled' as it would be in a gully.

Criterion 5 is whether the avalanche whirls through the air or flows along the ground. In the first case it would be the *Staublawinen* of the mountain people, best translated not literally as 'dust avalanche' but rather as 'airborne-powder avalanche'. The term for an avalanche which flowed along the ground would be a 'flowing avalanche'.* In practice it would seldom be used because any avalanche not described as an 'airborne powder' would obviously have flowed along the ground.

The old term *Grundlawinen* has not been incorporated into the classification as it is too vague, unless taken only to mean the massive wet-snow avalanches which come down well-defined tracks in spring bringing with them rock, soil and other debris. But even then the term lacks precision because such avalanches can originate in different ways, as will shortly be explained.

Beginning from the broad concept of criterion 1, that avalanches are either loose-snow or slab avalanches, it is convenient to apply

---

* It can be said that the movement in an airborne-powder avalanche is aerodynamic while that in a flowing avalanche is hydrodynamic.

the other criteria to them and so describe the various types of avalanche which exist. We shall not, however, bother with criterion 4 as it has little bearing on avalanche types; they may all come down unconfined slopes or be channelled in gullies; it is purely a matter of terrain.

LOOSE-SNOW AVALANCHES

As shown in the sketch in the classification system, these avalanches start from a single point and produce a pear- or tongue-shaped track. It is more usual for them to involve the top strata of the snow cover only so they are normally surface avalanches rather than full-depth. But, by criterion 3, they can be formed of either dry or wet snow.

Dry loose-snow avalanches. These avalanches can usually be seen in some profusion a few hours after new snow has fallen in calm conditions. On many of the steeper slopes a few crystals will have started to move, upsetting the equilibrium of the crystals below. Gradually gathering momentum they set other crystals in motion, so fanning out and forming the elongated pear-shaped track. These avalanches are not particularly dangerous and seldom travel at more than 20–30 m.p.h. A skier can easily set one moving without much risk to himself. But there may be a narrow dividing line between these dry loose-snow avalanches flowing along the ground and the fearsome clouds of an airborne-powder avalanche storming down the mountain-side at up to 200 m.p.h., annihilating everything they encounter.

Airborne-powder avalanches. These are the most devastating of all avalanches. We will cover them here under the category of dry loose-snow avalanches even if they can also—and quite often do—start as soft-slab avalanches. The important point is that by criterion 5 of the classification, which describes movement, they have become airborne rather than flowing along the ground. The destructive power of the blast which these airborne avalanches create can hardly be credited. As an example, photographs 2a and b—showing the damage caused by an airborne avalanche at Vinadi in the Lower

Engadine on February 18th, 1962—are worthy of careful examination.

The general photograph shows a part only of the greatest ever recorded damage to timber by an avalanche in Switzerland. The avalanche started high on the summits above the village of Vinadi and the swirling snow travelled 2⅓ miles through a vertical height difference of 6,500 feet, dividing into several streams as it went. It snapped off and uprooted between 240 and 250 acres of sturdy larch and pine forest on the Swiss side of the Inn Valley, and it also did considerable damage on the Austrian side. The forest was from 120 to 150 years old. The road was blocked over a distance of 1½ miles and the work of salvaging the timber continued for more than two years.

In the general photograph it can be seen how the blast of the far stream of the avalanche jumped across the valley floor and laid waste quite a large area of timber on the opposite slope (top centre in the photograph). In the detail photograph it can be seen that few trees escaped, and if they did their branches were stripped and their tops snapped off.

It is not surprising that such blast is capable of plucking up human beings and carrying them large distances. There is the case of seven forestry workers of Glarus, Switzerland, who, in 1900, were climbing up a steep slope when an airborne-powder avalanche rushed upon them from above. Six were smashed to death, but the seventh survived to describe how he was seized by the wind and flung through the air at tremendous speed, 'head sometimes upward, head sometimes downward like a leaf driven by a storm'. He was blasted by snow from all sides, could not see and scarcely breathe. He lost consciousness and came to, lying in deep snow which had broken his fall and limited his injuries to a few fractures. He thought his flight had lasted a mere split second but, in fact, he had been carried through a height difference of 2,200 feet and through a distance over the ground of more than half a mile.

The blast associated with such avalanches can also travel large distances. There is, for example, the story of a coach being blown across a stream, complete with coachman and horses, near the Flüela Hospice. The avalanche had come down so far away that the coachman had not even noticed it. This must have been a very large avalanche, of course, but that small ones too can generate powerful blast was shown by the Langen bus tragedy of December 22nd, 1952.

The bus was on its way to the ski-resorts of Zürs and Lech in Austria, and it was filled with happy people who were chattering gaily about the coming holiday and about the special beauty of Christmas in the mountains. Not far from Langen the bus started to cross a bridge over the Alfenz, a mountain torrent, when the passengers heard that swelling bass hum with whistling overtones, that eerie rushing and whining which is the voice of an airborne-powder avalanche in full flight. The next instant the bus was flung off the bridge and into the torrent.

The survivors, only 11 of the 35 in the bus, spoke of so much air pressure that they thought their lungs would burst. Yet the avalanche was less than 20 feet wide; the bus in front of the ill-fated vehicle and the car behind were unscathed. Though most of those killed died by drowning in the water of the torrent, or from injuries received in the 30-foot fall of the bus, the legend, best treated with reserve, is that some of the victims were killed by air pressure rupturing their lungs.

In the 1920s, people began to propound theories to explain the destructive wind blast generated by airborne-powder avalanches. It was already known that these avalanches travelled at great speed (although it was not known why this was so), and the most obvious idea was that the blast was formed by the frontal pressure of the moving mass creating a gale; but this alone was insufficient explanation. Then it was suggested that the moving avalanche expelled the air from the snow cover ahead of it. Another belief was that, like any object moving fast through air, an avalanche left an area of low pressure behind it, a partial vacuum into which the air then rushed. This air then overtook the avalanche as it slowed and produced the blast.

To explain the high speed of airborne-powder avalanches, Albert Schlumpf published a theory in 1936. The basis of his idea was that the head of the avalanche became pressurized due to the resistance of the air it was passing through. He stated that behind this pressurized head there would be an area of low pressure into which the snow further back would accelerate. It would strike the rear of the avalanche head and propel it forward so that the avalanche was in effect self-energizing.

1. Typical striated track of a wet-snow avalanche. The sides and bottom of he mass stop, or move very slowly, while the centre continues to flow. The rooves are cut by snow-boulders

2. A large slab avalanche that has run across the Weissfluhjoch–Parsennhütte ki-run at Davos. The run was closed and the avalanche released with an explosive harge thrown from the cable-car that crosses the slope

13. The fracture line of a very large slab avalanche
14. A slab avalanche involving the whole of a large slope (Gaudergrat on the Küblis run at Davos). Note that the avalanche has started below a small cornice

Schlumpf also believed that within the pressurized head of the avalanche there was a sharp temperature rise. Helped by the effect of air friction, and the natural temperature difference between the avalanche starting point and its destination in the valley, the total heat generated would be sufficient to melt some of the snow and turn it into steam, he stated. Sudden expansion when an object was struck would produce an explosion and scatter debris in a wide arc as the compressed steam and air fanned out. This was all very dramatic but somewhat hard to credit.

Then Professor Wagner, a friend of Paulcke's, put forward the basic idea behind today's theories. In 1938, he stated that each particle of snow in an airborne-powder avalanche is surrounded by a layer of air, so forming a suspension with a density perhaps ten times greater than that of normal air. The suspension then flows like a heavy gas with very little frictional resistance.

In 1954 and 1955, a Swiss called Rohrer published his theories about airborne-powder avalanches, theories which he backed up with the results of painstaking practical tests and observations in the field. There is little doubt in my mind that Rohrer's explanations are correct.

Moving along the same line of thought as had Professor Wagner in 1938, Rohrer applied knowledge of gas flow to airborne-powder avalanches. The basic facts are that, firstly, when a gas or suspension flows through the atmosphere it drags with it a mass of air because of friction between the air and the particles. The energy of the original particles is therefore spread to a much larger mass; the air is nothing more than a passive stream being dragged along. The greater the energy of the original particles, and the more there are of them, the greater the speed imparted to the passive air and the more air involved.

Secondly, when a gas stream passes through a pipe, friction along the walls of the pipe slows the outer layers of the gas while the centre of the stream flows much faster. Indeed the central core may flow at double the average speed of the whole stream. A circular movement of particles from the centre towards the outside is created, and experiments show that particles suspended in the stream alter their distribution, an effect first noticed by Isaac Newton. The higher the speed the greater the swirl from the centre to the outside. Turbulence is being created.

Both of these facts about gas behaviour fit neatly when applied to airborne-powder avalanches. A dense suspension of snow particles drags a large amount of passive air with it. As the suspension moves, friction along its outer surfaces, caused by the surrounding air which has the same effect as the pipe mentioned above, generates swirl and turbulence. In the case of an avalanche, too, any unevenness in its path will increase the turbulence, as does an irregularity in the wall of a pipe through which gas is flowing.

Rohrer proved the existence of these swirls by setting metal flags on rods in the path of an airborne-powder avalanche to act as telltales of the blast direction after the avalanche had passed. Some flags were left at an angle as large as 36° from the general line of descent of the avalanche. Further experiment seemed to show that the duration of these swirls or cross-gusts was in the region of one-tenth of a second only.

When it is remembered that these swirls move faster than the mass as a whole, and perhaps even twice as fast as a mass which may be travelling at 200 m.p.h., their great destructive power is made clear.

The swirls operate in the vertical as well as the horizontal plane of course, and it requires little imagination to visualize the effect when a house is struck by such short, sharp blasts from different sides. It simply disintegrates. It explodes into thousands of fragments which are carried in many directions. Rohrer's hypothesis explains therefore why debris is often left in a circular pattern, a fact which puzzled people for years and led to the belief that in some way the pressure of the avalanche filled the house and blew it outwards.

Rohrer himself has seen a 30-centimetre (12½-inch) thick reinforced concrete wall hit by an airborne-powder avalanche. The rapid blasts from varying directions blew the concrete out from between the reinforcing wires which were but slightly bent.

Rohrer's explanation of what happens *within* an airborne-powder avalanche in no way alters the fact that a pressure wave of snow-free air precedes it. This has been shown on film when trees have been seen to break off before the snow-cloud reached them. Nor does Rohrer's work preclude the fact that air is sucked into the wake of an avalanche from the sides and behind. There is a case, for example, of a skier being sucked into an avalanche when he was standing above its starting point. There are also cases of houses being damaged by avalanches which have passed nearby; the debris has been sucked *towards* the avalanche path.

Rohrer also carried out some interesting observations as to why a flowing, dry, loose-snow avalanche, moving harmlessly down a slope and making its pear-shaped track, should suddenly swirl into violent airborne movement and accelerate rapidly. This transition can obviously take place if the avalanche falls over rocks or is deflected upwards into the air by some ramp in its path, but Rohrer carried out his tests on an even slope.

Stripes of dye were sprayed on to the snow in the direction of the avalanche path, various gauges were set up and the avalanche released with explosives. The snow flowed like a normal loose-snow avalanche and there was little intermingling of the dyed snow from the stripes with the undyed snow. This proved that the flow had remained laminar with little turbulence. The average speed of this avalanche was only 17 m.p.h.

Two winters later conditions were again ripe for carrying out a test on the same slope, but there was more snow than there had been on the previous occasion. The avalanche was released and for 120 yards the flow remained laminar. Then, suddenly, turbulence set in and a snow-cloud developed. By the third speed-measuring station the avalanche was travelling at nearly 75 m.p.h.; all instruments lower down the slope were ripped out. The dyed snow was thoroughly intermingled with the rest and the avalanche left no heaps of piled snow in the valley, as it had on the previous test; rather it scattered fine powder over a wide area.

The conclusion to be drawn is that if an ordinary dry, loose-snow avalanche is big enough to reach a certain critical speed along the ground, the turbulence set up by friction along the bottom and sides of the snow mass will cause it to whirl into the air. It then becomes an airborne-powder avalanche. What the critical speed is remains uncertain, but it is perhaps significant that the slowest airborne-powder avalanche that Rohrer has timed was moving at 53 m.p.h. (the fastest was moving at 190 m.p.h.).

It must be added that not all dry, loose-snow avalanches are either fully airborne or flowing. As often as not they are a mixture of both types, part flowing along the ground while a cloud of snow dust swirls up and races ahead.

It would seem then that, thanks mainly to Rohrer, the mystery

of airborne-powder avalanches has been explained, but they remain
as devastating, fearful and beautiful to behold as ever.

Wet loose-snow avalanches. Wet loose-snow avalanches (see photograph 8) have the same shape as normal dry loose-snow avalanches,
that is to say they start from a single point and make a pear- or
tongue-shaped track. The commonest time for these avalanches to
form is in spring when the snow becomes saturated with melt water
which weakens the bonds between the snow crystals. As can be seen
quite clearly in the photograph these avalanches often start near
rocks. This is because the rocks, heated by the strong spring sunshine, cause even more thawing, hence melt water, in their immediate vicinity.

The snow in these avalanches tends to ball as it moves, forming
snow boulders among the debris. These, too, can be seen in the
photograph. The slushy snow in the small type of avalanche illustrated moves slowly, often at a mere 10–15 m.p.h., and a skier can
often avoid it. If unfortunate enough to be caught, however, the
danger is very real, for these wet snow avalanches solidify like concrete when they stop.

A few years ago the Parsenndienst received a message that a
woman skier was buried up to the ankles at the edge of a small, wet
loose-snow avalanche; but when the patrolmen reached the spot
there was nobody in sight. Then they noticed two holes in the snow,
and there was a ski-boot still attached to a ski at the bottom of each
hole. The snow had solidified so completely over her skis that the
woman had been unable to move them. Fortunately, her boots were
old and not properly tightened, so she was able to escape from them
and walk away on stockinged feet.

Occasionally, however, avalanches of loose, wet snow reach
immense proportions when they come thundering majestically down
gullies, bringing with them rocks and earth and piling enormous
heaps of dirty snow in the valley (see photograph 10).

These are the *Grundlawinen* of the mountain people. They
usually start in a large catchment area high on a mountain flank, are
funnelled down through a gully and fan out again when they reach
the valley floor. They sometimes run distances of 2 miles or more
and contain several hundred thousand tons of snow.

In 1925, A. Allix calculated conservative horsepower figures for these avalanches, using 160,000 tonnes (1 tonne = 2,204 pounds) as an average weight—about 200,000 cubic metres of wet snow. He took 2,000 metres (6,500 feet) as the average vertical drop through which they fall, and he gave them a low speed of 36 k.p.h. (22 m.p.h.). In such cases, 20 *million* horsepower are involved.

Using the greater figure of 2·5 million tonnes of snow, said to have been the weight of a mammoth avalanche which fell in Italy in 1885, some 300 *million horsepower* were generated. Quite recent estimates show that a million tonnes of snow can quite easily be contained in one large avalanche, so perhaps the estimate of 2·5 million tonnes for an isolated freak was not wildly exaggerated.

Whatever their cubic content, such avalanches, some more than 100 feet thick when they stop in the valley, have dammed rivers and closed roads so that tunnels have had to be pierced through the snow. In 1860 the River Isère in the French Alps was dammed in this way. The river cut a tunnel beneath the snow but the remainder served as a bridge for 17 months before collapsing. In 1876 such an avalanche blocked the road at Zernez in the Lower Engadine and a tunnel was driven through it. Despite a hot summer it was 14 months before the last traces of the avalanche had melted.

Colossal as they are, these avalanches seldom do much harm to life and property: their tracks are too well known and some come down regularly every spring. They have been given names and are a feature of their area, but woe betide anyone who does get caught in one. The heavy, wet snow contains little air to breathe and the trees and rocks they contain make dangerous fellow travellers.

The impression may have been given that wet loose-snow avalanches only take place in spring, but, of course, they can occur whenever thaw conditions prevail. They can be produced in the depths of winter, in new or old snow, by warm weather or rain.

A final feature of these avalanches which should be mentioned is the curious grooved tracks they sometimes form (see photograph 11). This is caused by the sides and bottom of the avalanche being slowed to such an extent or even stopped, by friction, that they build walls and a slide path down which the centre of the avalanche then travels, the snow boulders cutting grooves and striations as they go.

SLAB AVALANCHES

Returning to the first criterion of the classification, we find the other broad category of avalanches, the slab type, which, in contrast to loose-snow avalanches, are made up of cohesive snow. Slab avalanches break away from a whole area all at once, rather than starting from a single point, and they leave a fracture line which is often jagged, or zig-zag as shown in the classification diagram (see photographs 7, 9, 13, 14).

The name slab avalanche is derived from the fact that the snow moves away as a slab, breaking up as it goes into smaller slabs, or even fine powdery debris according to the snow's original inner cohesion, how far it slides, and whether the slide path is smooth or uneven.

Any snow layers with a certain degree of inner cohesion which are lying on fragile strata such as surface hoar or cup crystals, or which are lying on a base to which they are not firmly attached, can form a slab avalanche.

The snow may be very cohesive as in the case of hard wind-packed snow, or it may be only slightly toughened by the effect of wind, or it may have obtained its cohesion through the natural process of settling and firming. But as long as the avalanche breaks away leaving a fracture line, this classification system recognizes it as a slab avalanche. This is in contrast to many earlier classifications which reserved the term slab avalanche for avalanches of very hard wind-packed snow or 'wind slab'.

However, given that slab avalanches may consist of very hard or only slightly toughened snow, it became the practice in the U.S.A. in the early 1960s to differentiate between 'hard-slab' and 'soft-slab' avalanches. Following the 1965 International Avalanche Symposium in Davos, it was agreed to incorporate this distinction into the classification system.

Taking the criteria of the classification in turn we shall see what bearing it has on slab avalanches. By doing this we shall discover what forms these avalanches take, just as we did with those of loose snow.

By criterion 2, slab avalanches can either be full-depth or surface avalanches according to whether the whole snow cover had moved or a few top strata only. Surface slab avalanches are the more usual with a fragile stratum in the snow cover as the horizontal break-off point and slide path. Nevertheless, the whole snow cover may slide, especially when the ground is smooth. Grassy slopes are the usual scene for such avalanches, and this is particularly so if the grass has not been cut for hay in the previous summer. The long stems lying downhill form a slippery slide path.

By criterion 3, slab avalanches can either be dry or wet:

Dry-slab avalanches. By far the commonest form of dry-slab avalanche is caused by wind packing of the snow during the blizzard, or even wind blowing the snow about and packing it after it has fallen.

Precisely where the wind deposits these masses of hardened snow was a subject of debate for many years. Some people insisted that wind-slabs formed on windward (exposed) slopes, while others stated that they formed on lee (sheltered) slopes. Others maintained that they formed on both windward and lee slopes. Paulcke was among those who claimed that they were to be found on both, but he differentiated between 'wind-pressed' snow on windward slopes and 'wind-packed' on lee. Other writers made other distinctions, and confusion reigned.

To a certain extent these experts were at cross-purposes. Certainly, wind-hardened snow is to be found on windward slopes and ridges, but usually in too small a depth to form an avalanche. The deep slabs, anything from 1 foot to 12 feet thick, which constitute an avalanche menace, occur mainly on lee slopes, often below a cornice as shown in the last chapter. It is, however, very important to realize that *wind-packed snow will accumulate in any place where there is shelter from the wind. It is therefore quite possible for wind-slabs to occur in depressions or gullies, or near rock outcrops on what is basically a windward slope.*

One of the characteristics of wind-packed snow is that it does not settle much. The fine dendrites have been so damaged already that destructive metamorphism has no part to play in rounding the crystals. But the snow strata beneath a wind-packed layer may still be settling and a gap may develop under a slab. The slab may then lie there, its surface as hard as concrete but its foundation as fragile as glass. The creep of the snow cover puts tension forces on the slab and it may remain in place for weeks on end as the neatest booby trap that Nature can devise.

The hard surface of the slab inveigles the unwary into thinking himself on safe snow, but the trap has a hair trigger. The slightest disturbance can spring it, whereupon, with a noise which varies from a click to a loud detonation, hundreds or thousands of tons of snow explode into violent downhill motion.

It is this type of avalanche which claims most lives among skiers, and an accident occurred on February 8th, 1964 which is an example of the malevolent way they can lie in wait for the unwary.

It was a fine day in the Swiss Alps though the extreme shortage of snow in the ski-resorts was making people wish for a blizzard rather than sunshine. Apart from about 5 inches of snow which had fallen over a week before, accompanied by strong wind, it had not snowed since early winter. However, for a 21-year-old German, Detlev Döring, who was working as a cook in a Klosters hotel, the skiing conditions were good enough. He was virtually a beginner and he was going to make the most of his free afternoon in the sunshine, and clear air.

He was on the Kalbersäss run above Klosters and was making his way steadily downwards, though his progress was punctuated by regular falls. Then, in one fall his safety bindings released a ski. The retaining lanyard round his ankle broke, and Döring sat in the snow and watched in dismay as his ski trundled out of sight over a bump. He set out after it with difficulty and then found it easier to remove the other ski and go on foot. A ski-instructor nearby, seeing him leave the run, remembered the notices in the cable-car stations: 'Danger of Snow Slides on Steep Slopes'.

'Don't go down there!' the instructor shouted.

But Döring probably had no inkling of why the instructor had shouted this, and he called over his shoulder:

'I'm going to fetch my ski.'

He struggled on to the lip over which his ski had disappeared and looked down into a steep but shallow gully some 40 yards wide. The line made by his ski was traced lightly down the otherwise virgin surface.

In all probability without a second thought, he stepped down over the lip. There was a muffled detonation, like an underground explosion, and cracks ran like lightning across the snow around him. In an instant the surging snow had him in its grasp and he was struggling for his life.

The Parsenndienst rushed to his rescue, but when they located him after an hour and a half they found that they had once again lost their race with death. After an hour or so of revival attempts the young man's body was sadly loaded on to a sledge for transport to the valley, yet another victim of a dry-slab avalanche.

These avalanches, whose original cause is wind-packing, can be very small yet still lethal to skiers. The one which killed the unfortunate Döring was of moderate size—about 40 yards wide and nearly 200 long—but a number of skiers have died in avalanches with a width of only 20 yards and a length of less than 100. Indeed, one of about these dimensions killed another skier in the Parsenn area just three weeks after the Döring accident.

It is quite possible that the apparent harmlessness of these small, dry-slab avalanches is the fundamental reason for the lack of respect accorded them by skiers; yet once a skier has been surprised by that detonation near him, and seen the rapidity with which the snow leaps, he will never lack respect again. Even the smallest avalanche of this type can produce a pile of snow 7–8 feet deep at the end of its run, and many are the dead dug out from under a mere 2–3 feet of avalanche snow. On the other hand, of course, these avalanches can be several hundred yards across and involve a whole slope (see photograph 14). Only a fool could then fail to recognize their threat.

Wet-slab avalanches. Once in motion, wet-slab avalanches are very similar to wet loose-snow avalanches. They differ only in the form of the break-away, starting from a whole area rather than a point, as already explained.

Melt water in warm weather is the most usual cause of wet-slab avalanches. The water percolates down through the snow until it reaches either the ground or an impervious layer such as ice. The water then flows along this layer and breaks the bond between it and the snow above. It produces a full-depth avalanche if the break occurs along the ground, or a surface avalanche if only the snow above an ice stratum moves.

Wet-slab avalanches have a characteristic form of motion for the first few yards of their run. A crack appears in the snow cover and then, quite slowly, the cleft yawns open. Below this rupture, the snow, which is of course warm and plastic, forms waves and creases as it begins to move. It may heave itself up into folds like an enormous cloth or carpet. Then, once it has gathered speed, the snow slab breaks up and the movement becomes similar to that of a wet loose-snow avalanche, with the typical formation of snowballs and boulders.

It is not uncommon to see the early stages of one of these avalanches which has failed to develop momentum. The snow cover may split open, especially on south slopes in spring; the folds of an incipient wet-slab avalanche form, but the movement then ceases, perhaps owing to the snow cover being firmly anchored lower on the slope, or to the slope being too short to avalanche properly. The initial behaviour of wet slabs has led to them being termed 'snow-cloths' or 'snow-carpets' in the past.

It is quite possible for these wet-slab avalanches to form the great *Grundlawinen* described in the section on wet loose-snow avalanches, hence the earlier remark that the name *Grundlawinen* was imprecise as they could originate in different ways. For a wet-slab avalanche to form a *Grundlawine* of the mighty proportions described earlier it would have to start in a large catchment area and release a lot of other wet snow as it went.

Returning once more to slab avalanches in general and how the rest of the classification system applies to them, it should be made clear that, in the interests of simplicity, certain rather arbitrary distinctions have been made in the preceding pages. For example, airborne-powder avalanches were placed under the category of dry loose-snow avalanches with the qualification that they can also start as soft-slabs. However, even a hard-slab avalanche could, in theory, become an airborne-powder avalanche if it fell over rocks or such uneven terrain that its originally cohesive snow pulverized and whirled into the air. A dry-slab avalanche beginning in the shade could run into a sunny zone and collect wet loose snow. An airborne-powder avalanche may not be entirely airborne, or a dry soft-slab may, and usually does, have some snow-dust in the air above it. Thus, avalanches may sometimes be mixtures of various types and be difficult to classify precisely. Nevertheless, the classification system of avalanches in their pure state is important as a basis for communication and explanation.

GLACIER AVALANCHES

Glacier or ice avalanches are in a category of their own, some-what dull cousins of those enigmatic snow avalanches whose complexities are still not entirely solved. It is a fundamental fact that glacier ice in itself is an inert, unvarying material, and a glacier avalanche is therefore more akin to a stone-fall than a snow avalanche. It is not that ice avalanches lack in spectacular appeal, as anyone who has spent any time at Kleine Scheidegg above Wengen in the Bernese Oberland will know. Streams of ice-blocks the size of houses frequently hurtle down the Jungfrau massif with a thunder that makes the ground tremble.

Nevertheless, ice avalanches lack that peculiar fascination of snow avalanches. They are not controlled by that intricate gamut of factors like weather, wind, terrain, snow metamorphism, altitude, temperature and so on (all to be covered in the next chapter).

Ice avalanches are almost invariably the simple result of glacier movement; the ice moves slowly towards the edge of a drop until it falls over it.

Ice avalanches have none the less caused great catastrophes, for, though the majority of them are confined to the high-altitude areas where they affect only mountaineers, a few hanging glaciers do menace inhabited valleys. Part of the Mattertal for instance, the valley at the head of which lies Zermatt, is threatened by the Bies glacier on the Weisshorn.

The village of Randa, below Zermatt, has been severely damaged twice, once in 1737 and again in 1819. On the latter occasion the ice became so pulverized that the effect was of an airborne-powder avalanche—yet another example of an avalanche confounding all classification systems. The blast 'threw the house and roof timbers about like straw', depositing them as much as a mile away. In 1972 the hanging glacier was again large and menacing and Institute staff were called in to advise. On the basis of theoretical calculations, tests with models and the study of past disasters at Randa, they were able to develop several hypotheses of what would happen when the ice fell. And in fact, when only part of the hanging glacier did fall, in August 1973, it obligingly fulfilled their most favourable prediction and caused no damage.

The villages of Bionnay, St. Gervais-les-Bains and Fayet, in the

Haute Savoie, were ravaged by a muddy torrent at 1 a.m. on July 12th, 1892. It carried away more than 30 houses and claimed about 150 human lives. What had caused the muddy torrent was not at once apparent, but an investigation was carried out by F. A. Forel who summed up his report by writing: 'Three days after the catastrophe I ascertained that it was due to an avalanche from the hanging glacier of Têtes Rousses. The ice avalanche, after having dropped 1,500 metres (5,000 feet) in the first part of its course, which was 2 km (1¼ miles) long, changed from ice to a muddy, semi-liquid and flowed a further 11 km (7 miles) down a slope of 1 in 10. With a total drop of 2,500 metres (8,250 feet) and a total length of 13 km (over 8 miles) it was one of the greatest known examples of this phenomenon of nature.'

It remained one of the greatest known of these phenomena until January 10th, 1962, when the Huascaràn avalanche took place in the Santa Valley, Peru. This is the largest single avalanche disaster on record. It killed more than 4,000 people, 10,000 animals and completely destroyed six villages, and partly destroyed three others. A young Peruvian geologist, Benjamin Morales, carried out an investigation nine days after the disaster and presented a fascinating paper on the subject during the Avalanche Symposium in 1965. Morales's description of this stupendous phenomenon would be incredible were it not so fully documented.

North Huascaràn, at over 22,000 feet, is the second highest mountain in South America and its summit is covered by an ice cap. The avalanche began at an altitude of about 21,000 feet, when part of the ice cap broke away, and it finished in the Santa Valley at 8,000 feet, so travelling through a vertical height difference of 13,000 feet. The distance it covered was 10 miles and it did this in 15 minutes. (A telephonist in a village not destroyed saw the avalanche start and timed its course.)

A mass estimated at between 2·5 and 3 million cubic metres broke away along a front about half a mile in length, and where the ice cliff at the edge of the cap is about 180 feet high. In the first part of its course the ice plunged more than 3,000 feet down a near-vertical rock face, pulling out immense blocks of rock as it went. The mass landed in the glacier cirque below with tremendous impact; the noise was heard for many miles around and an enormous cloud of powdered ice swirled up and blotted out the summit of

the mountain. The concussion started a secondary avalanche of
about 300,000 cubic metres which followed the main one down.
The main avalanche also tore out more ice in the glacier cirque, and
by the time it reached the glacier tongue its volume had swelled to
about 5 million cubic metres of ice and rock. The mass moved
down the 35° gradient of the glacier at about 65 m.p.h. and then
tore into the twisting channel between the lateral moraines.

At the first bend, it climbed more than 500 feet to the outside rim
and deposited a large quantity of material outside the channel. The
avalanche took with it, as it went, more and more of the moraine until
it formed a rolling mass more than 175 feet thick. At the second bend
in the channel part of the avalanche climbed 350 feet on to a small
plateau and left blocks of rock measuring up to 50 feet by 25 by 30.

Thereafter, the ravine had a gradient of only 8–10° but the avalanche rushed on, gouging out both slopes of the channel and
destroying a village. At a slight curve further down, a branch of the
avalanche climbed a hill 275 feet high, left a 6,000-ton boulder on
the top, and then rejoined the main stream. The avalanche then
swerved into a larger valley, climbed 265 feet up one side and
destroyed another village. Further on, where the valley widens and
joins the main Santa Valley, the mass began to fan out. It spread to
a width of 1¼ miles, annihilated four more villages and partly destroyed two others. It reached the Santa River travelling at about 20
m.p.h. and climbed 100 feet up the far bank to destroy part of yet
another village. The avalanche eroded so much rock, soil and sand
during the course of its 10-mile run that its final volume is estimated
to have reached the almost unimaginable total of 13 million cubic
metres.

Few people in the path of the avalanche were able to escape. One
who did was the telephonist in Ranrahirca, the largest of the villages
destroyed. She was warned by a colleague in a nearby village who saw
the start of the avalanche. But she thought the warning was a joke—
and one in poor taste at that—before she took it seriously and fled.

Another survivor was a potato merchant who was waiting for a
lorry by the roadside near Ranrahirca. He heard a noise 'like many
aeroplanes flying over Huascarán' and, looking up, saw the avalanche already well on its way. He began to run for safety and shouted warnings to others as he went. Many people paid no attention,
and two women even remarked that 'the snow of Huascarán always

fell like that'. Others ran to their houses or into the churches where, of course, they were killed. The potato merchant was almost exhausted when he saw some horses tethered by the roadside. He hauled himself on to one and galloped away. Looking back, he saw the avalanche cross behind him. At the front of the mass the ice blocks were constantly cracking and breaking up, creating a pall of powdered ice. He saw the village of Ranrahirca engulfed.

The Santa River was dammed but, fortunately, the water soon burst through the debris and the flooding down the valley was not as severe as it might have been. Even so, roads and bridges were destroyed as much as 28 miles away.

Flooding is often a by-product of ice avalanches. The Rhone valley above the lake of Geneva has been flooded on several occasions when tributaries have been dammed. In 1595, an ice avalanche blocked the River Dranse, and when the dam burst 140 people lost their lives. The same happened in 1818 when the lake formed was 1½ miles long, 200 yards wide and 200 feet deep. When the ice wall collapsed, it was calculated by an engineer called Escher that the water flow was 300,000 cubic feet a second—'greater than the flow of the Rhine below Basle'. Bodies were found at Vevey, more than 30 miles away.

More recently there occurred the much-publicized disaster at Mattmark, Valais. A hydro-electric dam was being built, and work had been in progress for five years. But on August 30th, 1965, when a further two months' work would have seen the scheme complete, more than a million cubic metres of ice broke from the Allalin glacier and killed 88 men in and around some maintenance buildings. It was most unfortunate that the avalanche fell just when work-shifts were being changed, and double the usual number of men were near the glacier.

The cause of the avalanche was probably the wet summer, during which rain-water gathered in the crevasses and lubricated the undersurface of the ice. Many people have said that the avalanche could have been foreseen, but this is hardly true. On the very day that it occurred men had worked near what was to become the avalanche fracture-line without noticing anything unusual.

It would seem, however, that since the 1850s, when the glaciers began a slow retreat, the chances of ice avalanches invading inhabited valleys have been diminishing also.

6
Build Up and Release

The creation of a potential avalanche and its eventual release depend upon a number of interrelated factors—so interrelated and complex in fact that it is still impossible to forecast with any exactitude when an avalanche will start and what it will do. Most people who have not studied the subject believe that, *a priori*, the steepness of the slope is the most important factor; but in reality the properties of the snow cover and the quantity of snow are more significant.

For those avalanches which occur during or immediately after a snowstorm, the amount of snow deposited plays an important role. The following is a very approximate guide to the relationship between depth of new snow and the avalanche danger it constitutes: 6 inches to 1 foot—slight local danger for skiers; 1 foot to 2 feet—considerable local danger for skiers; 2 feet to 3 feet—local danger for roads and communications; 3 feet to 4 feet—onset of great general danger, including for exposed houses; new snow in excess of 4 feet threatens whole villages with destruction.

These figures can only be approximate because, as the saying goes, it is not only the quantity of snow which falls that causes avalanches, but also the manner in which it falls. For example a fall of only 4 inches of snow accompanied by high winds can produce local dry-slab avalanche conditions which are dangerous for skiers. In fact, wind nearly always increases the danger, especially winds with speeds between 20 and 40 m.p.h. It is not known with certainty, but there appears to be an upper limit to wind-speed above which the danger-increasing effect begins to tail off again, probably in the region of 60 m.p.h. This could be because such very high winds hardly allow snow to settle anywhere, except in almost flat and totally sheltered places.

The intensity of the snowstorm is also very important. If a fall of say 30 inches is spread over a long time it produces far less likelihood of avalanches than if it falls over a short period. This is because the continuous settling of the snow under the influence of destructive metamorphism has a better chance of keeping pace with a gentle snow-fall and so stabilizing the layer. Settlement is faster at high temperatures so, usually, the warmer it is during a snow-fall the shorter the danger period will be. These facts explain why a break in a snowstorm often removes much of the avalanche threat, and why a drop in temperature during a snowstorm *may* increase it. (A drop in temperature during a snowstorm will *not* increase the danger level if its effect on crystal type and snow density has, in balance, a strengthening effect on the new snow cover, even if it is not settling fast. This is just one of those imponderables that add to the fascination of the study of snow and avalanches.) Also very important, of course, is the stability of the existing snow cover to which the new fall is being added.

All in all then, one should think in terms of avalanche weather and times rather than in terms of avalanche terrain, though this is not meant to imply that terrain has no bearing on the matter.

THE INFLUENCE OF TERRAIN IN AVALANCHE CREATION

Steepness. Avalanches can occur on slopes down to 22° and in special circumstances on slopes of even less gradient. The figure of 22° may seem somewhat arbitrary but it is derived from the friction angles of granular snow. Static and kinetic friction angles apply to all loose granular materials and a brief explanation is necessary here. The static friction angle of any given granules is the angle on which they will remain without moving, held there by the static friction between the particles. An 18th-century scientist called Coulomb demonstrated that this angle *generally* corresponds to the natural shape taken up by a heap of the material in question. This must be qualified by the word 'generally' because the static friction angle also varies according to the weight placed on the particles.

Once a granular material is in motion, however, it becomes subject to the smaller force of kinetic friction. The kinetic friction angle of a material is the angle on which the material will move once set in motion. Therefore, from the kinetic friction angles of the various types of snow we know the minimum gradient on which an avalanche of that snow can be started. Of all the types of snow in their natural state, spherical granules have the lowest kinetic friction angle at 22–23°. For angular crystals it is rather higher, reaching 35° for cup crystals.

The static friction angles of snow are confused to some extent by the fact that snow crystals lying in the snow cover are often aggregated and do not, therefore, behave as a true granular material. It follows that static friction angles are only of importance for crystals that do not form bonds. Thus for cup crystals the static friction angle is about 45°, and for powder snow, before it becomes aggregated, the angle lies between 50–60°; but one must bear in mind variations brought about by pressure on the crystals from the weight of snow above.

The delicate, dendritic crystals of new snow are a special case when considering static and kinetic friction angles. These crystals, although they do not form bonds in the true sense, interlock by means of their fine branches and needles. So effectively do they mesh together that, as long as the layer is not too thick and heavy, they can hang on the vertical. This interlocking, in effect produces a sort of aggregate, but it is usual to say that the static friction angle of new snow can be up to 90°.

In this way, a layer of new snow which is not too thick forms a very stable mass. But, if these dendritic crystals receive sufficient shock, or are set in motion with sufficient impetus, they will fracture into fine fragments. This snow-dust has the lowest kinetic friction angle of all snow, a mere 17°. Fortunately, a layer of dendritic crystals is so plastic and loosely knit that any shocks which the snow receives are normally absorbed without the fracturing of the crystals taking place. In fact, avalanches of pulverized new snow on 17° gradients are sufficiently rare to be disregarded, and so it is usual to give the minimum slope angle for an avalanche of dry snow as 22°.

It is not possible to be so categorical about wet-snow avalanches. Reliable observers have seen them start on gradients of only 6–7°; but in these cases there is so much free water in the snow that it behaves like a liquid rather than a granular substance. Avalanches of this type move so slowly that they cannot be considered a menace.

Once started, certain avalanches may continue to move down very shallow gradients, even without the effect of accumulated momentum which can take an avalanche a considerable distance over level or even rising ground. Two types of avalanche may move on these shallow gradients: the first is a wet-snow avalanche (but not necessarily one so wet that it flows like water). In this case, the front and the sides of the mass create a smooth path, lubricated by moisture, down which the central and rear parts of the avalanche slide easily. The other type of avalanche is, of course, the airborne-powder because its movement is subject to so little frictional resistance.

Airborne-powder avalanches can retain a great deal of destructive power even after travelling considerable distances down shallow gradients, and a good example of an avalanche which behaves in this way is the one in the Schiatobel at Davos. This avalanche is known to have come down several times since Davos was built and it did so again in 1962.

The part of the town known as Davos-Dorf is dominated by a peak called the Schiahorn whose flanks, especially the south and east, have spawned several large and destructive avalanches within recorded time. One of the most devastating of all occurred at Christmas, 1919. Several people lost their lives and there was considerable damage to property, including to a large sanatorium which is now the Derby Hotel. In its dying convulsions, the avalanche burst through the rear windows of the sanatorium dining-room and out of the front, piling a snow-and-furniture mixture all over the terrace. Fortunately no one was in the room because lunch had just been finished. The avalanche of 1919 persuaded the authorities to build the anti-avalanche fences now visible on the southerly slopes of the Schiahorn, and after their completion peace reigned, apart from one avalanche in 1935.

But the Schiahorn avalanches were not to be tamed quite so easily. They attacked again in 1962, this time down the south-west and west flanks where the defences were less comprehensive. Along the foot of these slopes there runs a deep and narrow ravine, the Schiatobel, which leads into the Davos valley. The total length of the ravine is about $1\frac{1}{4}$ miles, but its average declivity is only 15°.

At 06.00 hours, on February 17th, 1962, some 60,000 tonnes of snow plummeted down the Schiahorn westerly flank and slammed into the Schiatobel. In the ravine, the snow was diverted through about 80° and began its mile-long journey towards Davos. It rushed zig-zagging down the gorge, ripping out trees on either side as it went. Much of the snow was left behind but, nevertheless, more than 8,000 tonnes were still moving fast enough to tear down the final trees and burst into the Davos valley. A bridge and four houses were severely damaged, but saddest of all was that a 9-year-old boy, Andreas Eitle, was killed by the snow that buried him in his bed.

In some places the gradient of the Schiatobel is only 11–12°, and it is remarkable that the avalanche can be so destructive after flowing more than a mile down such a twisting track. Admittedly, the slopes on which the avalanche originates have a gradient of 39°, and it must gather a great deal of momentum in its early stages; but much of this momentum must be lost when the avalanche makes its near right-angle turn to flow down the Schiatobel. It is thought that one factor may contribute to the behaviour of this avalanche: as the snow/air suspension rushes down into the ravine it is compressed, and it may even compress the air already in the bottom of the ravine. The violent expansion along the gorge would then help to carry the avalanche, adding considerably to its velocity and destructive power. Certainly, after the 1962 avalanche, the effects of air blast were well in evidence: pine needles and small branches were scattered over a wide area.

In passing it must be mentioned that the Schiatobel avalanche behaved in almost exactly the same way in 1919, but it was the other avalanche from the south-east Schiahorn flank that did the damage and tore through the sanatorium. For this reason, the defences against a recurrence of the Schiatobel avalanche were not granted the same importance and proved ineffective in 1962. Needless to say, the error has since been rectified.

The ability of avalanches to continue their movement down shallow gradients must be borne in mind when choosing a route on a ski-tour. Nor must it be forgotten that, given the right conditions, any slope between 22° and 50° *can* become an 'avalanche slope', in the sense that an avalanche may actually start on it. This does not, however, alter the fact that on steeper slopes avalanches are more common, because on such slopes the 'right conditions' are more easily satisfied. The most devastating avalanches usually start on slopes between 30° and 40°. Over 50° avalanches are rare because insufficient snow ever accumulates: it merely slides off repeatedly during a snow-storm, in small quantities at a time and without constituting a danger.

Configuration and features of terrain. For an avalanche to occur there must, of course, be a loss of equilibrium in the snow cover. In the case of a loose-snow avalanche this can be brought about by the toppling of a single crystal which begins a chain reaction; but in the case of a slab avalanche it can only follow a fracture in the snow cover. As soon, therefore, as snow has any cohesion and strength, factors creating stress which could lead to a fracture are of great importance. One such factor is the vertical slope profile (see Fig. 6). Obviously, the stresses created by creep on convex slopes make such slopes more likely to avalanche than concave ones.

Variations in horizontal slope profile, forming gullies and ridges, also create stresses as the snow creeps from the shoulders into the hollows, but these stresses are less avalanche-provoking than those caused by variations in vertical slope profile. Nevertheless, depressions and channels down a slope are very often the scene of avalanches, mainly because they collect more than their share of snow, almost irrespective of wind direction.

Where several gullies interconnect and terminate in a common outlet, an avalanche down one frequently releases avalanches in the others. The movement of snow in the common outlet destroys the support of the gullies above. Photograph 15 shows such a case. There were five avalanches into the one gully, of which three are visible in the photograph. The first avalanche, top centre, was released by an American skier, Richard Muhl of Chicago, and he was carried away by it. The rescue team of 90 men finally located his body two hours later, at a point just beyond the bottom right corner of the photograph. He was under 8 feet of hard-packed snow. Such gully systems are particularly dangerous because anyone caught in one avalanche and covered by successive ones has virtually no chance of survival. It was extremely fortunate that Muhl's friends had waited to watch him ski the slope or they too might have died. This accident, incidentally, was another occasion when skiers had left the marked run despite warnings not to do so.

Slopes which have irregular surfaces, for instance terracing caused by underlying rock strata, or even terracing caused by the constant treading of livestock over the centuries, are safer than smooth slopes, unless the irregularities have been levelled out by snow. This may well be the case by late winter and then the terracing will only prevent the formation of full-depth avalanches. The more common surface avalanches will be unimpeded.

Steep rocks at the top of a slope, cliffs, or even trees render the slope more avalanche-prone because a pebble, an icicle or a little snow falling on the slope can be sufficient to set an avalanche in motion. And the point has already been made that the extra thaw effect around rocks in spring can be the cause of wet-snow avalanches.

Slope orientation. Most people know that a rise in temperature can produce avalanches, and the mechanics of this will shortly be explained. For the moment, it is sufficient to point out that there exists a relationship between temperature and avalanches and that the effect of sunshine on a slope is therefore important.

After a snow-fall, the slopes which catch the sun often avalanche within a few hours; but if they do not do so the continued effect of the sun produces a stable snow layer because it speeds the process of settling. In spring, though, the strong sunshine will cause these southerly slopes to shed many wet-snow avalanches as thaw water saturates the snow cover.

Slopes which get little or no sun remain unstable for longer after a new snow-fall because the snow settles and stabilizes more slowly at low temperatures. Such northerly slopes can be very dangerous if there is suddenly a general rise in temperature of the sort brought about by a warm wind like the Föhn. The accumulated snow which has neither avalanched, nor settled properly under the influence of sunshine, loses what stability it has. There have been many avalanche accidents when skiers have been skiing northerly slopes in safety and failed to notice a sudden rise in temperature; their next run down the same slope has proved fatal. Just such a situation caused the death of Mrs. Ulla Neave on the north-facing Drostobel at Klosters in March 1963. In that particular instance, it was not only the sudden and considerable temperature rise that her ski-instructor failed to notice: he also missed the 'Run Closed' sign.

Altitude. In the European Alps the majority of avalanches begin their journey from altitudes between 6,000 and 9,000 feet. Bearing in mind the fact that snow-fall in the Alps increases with altitude, it is perhaps surprising that winter avalanches are comparatively rare above 10,000 feet. But the slopes above this altitude accumulate little snow, partly because they are generally very steep, and also because of the type of snow crystal usually deposited at high altitude

—the crystals are of the basic, simple types, not the branching structures that cling together and can therefore lie on steep ground. And another factor is that the winds at high altitude are often in excess of 60 m.p.h., the speed above which it is thought that the formation of slab diminishes.

When Professor Haefeli was deciding where to establish the Avalanche Research Institute, either at the Weissflujoch, at an altitude of less than 9,000 feet, or at the Jungfraujoch, at well over 11,000 feet, he organized parallel observations at both places for a period of six months. His interesting conclusion was that the Jungfraujoch was too extreme to be suitable for avalanche research. After his early work in Davos, at an altitude of only 5,000 feet, he had had a craving to go higher, but he soon found that the Weissflujoch was high enough. Below 6,000 feet, avalanches become rarer for three reasons: less snowfall, more forest and higher average temperatures, which help the snow to settle.

Ground cover. Mountain people frequently talk of the value of knowing a slope in summer when its proneness to avalanche is being assessed. So far as full-depth avalanches are concerned this is a reasonable assertion. For example, if it is known that a grass slope which is usually cut for hay has not been cut in a given summer, it can be assumed that a full-depth avalanche is more likely to take place during the following winter, for instead of the grass stubble to anchor the snow cover, there is long grass, the most slippery of all ground surfaces.

Bushes and shrubs also help to anchor the snow cover but once they are completely buried there can be no guarantee against surface avalanches. In point of fact, shrubs can have a detrimental effect on the stability of the snow cover once they are buried. There are often spaces around the branches and twigs buried in the snow, and these spaces help air to circulate. This in turn leads to a more rapid rate of constructive metamorphism and the cup crystals so produced of course reduce the strength of the snow. In addition, the springy branches of certain types of evergreen can create tensions in the snow cover. Photograph 15, as well as illustrating several avalanches in a gully network, also shows that they have broken away in an area of evergreen shrubs.

Woodland provides the greatest protection of all against avalanches, except when there is a long and open slope above on which an avalanche can gather sufficient momentum and power to fell the forest. Flowing loose-snow avalanches, especially those of 'wild snow' sometimes actually start in a wood and flow down among the trees, doing little or no damage. A skier can be endangered by them, however, in that he may be flung against a tree. Such an event was one of Professor Haefeli's introductory experiences to avalanches in his youth.

Despite this, woodland can usually be considered avalanche-safe terrain *but the woodland must be dense*. The gravest of all single avalanche accidents involving tourists in Switzerland took place on a slope studded with small trees. This was the accident above Lenzerheide on February 10th, 1961, involving a party of schoolchildren from Glarus.

During the winter it is general practice for Swiss schools to have a 'Sport Week' during which parties of children go to mountain huts to ski. This week is in fact the most keenly awaited and best enjoyed holiday the children have. From February 5th, 1961, a group of 13 girls and 14 boys from the Canton School of Glarus had been in the Rascheinas ski-hut above Lenzerheide, under the leadership of a mistress from the school who held a ski-instructor's certificate. She was being assisted by Johann Jenny, a 20-year-old engineering student.

The group was a happy one and the first few days were enjoyable despite the bad weather and frequent blizzards. The north-westerly winds were biting cold and the driving snow stung their faces, but even this hardly dulled their enthusiasm as they set out each day for a short ski-tour in the partly wooded area above the hut.

But had there been a serviceable radio in the hut they would have known that extra Avalanche-Warning Bulletins from the Research Institute were broadcast on February 5th, 6th and 7th. Each bulletin ended with a sentence earnestly advising that all ski-tours be abandoned, except in the safest of terrain. Even had the leaders of the group heard the warnings their plans might not have been changed; they believed that the partial timbering of the slopes above the hut made them safe.

During the night of February 9th, the wind roared outside the hut with even greater intensity, but the leaders attached no significance to it. They planned to repeat the tour of the previous days. By morning the weather had improved a little and the party climbed up through the woods until they reached the flat eastern shoulder of the Crap la Pala. They took off their climbing skins and got ready for the descent of the Porclas slope. This slope faces east and is at an altitude of about 1,900 metres (6,250 feet). The few trees on the slope are no impediment whatever to avalanches. In fact, the Avalanche-Warning Bulletin issued that day could have been written with the Porclas slope in mind.

'The intermittent precipitation of the last three days,' ran the Bulletin, 'has produced snow-falls of 20–40 cm at altitudes above 1,500 metres. Strong westerly winds of long duration have led to widespread drifting of snow with resultant accumulation of large masses in eastward orientated catchment areas. Avalanches may well start from such areas and reach the valley floor.

'In addition, the very great slab avalanche danger to skiers has in no way diminished. *The danger is acute at altitudes above 1,800 metres and especially so on eastward facing slopes where large and treacherous deposits of snow are to be found.** No ski-tours should be undertaken other than in the safest terrain.'

From the top of the Porclas Johann Jenny started down with the group of better skiers, but 14-year-old Ruth Landholt fell and could not extricate herself from the deep snow. Johann Jenny went back to help her. The schoolmistress was passing just above with her group and she saw Jenny take off one of Ruth's skis and lift her to her feet. The schoolmistress skied on until she came to a clump of pine trees, where she stopped and turned to watch the progress of her pupils across the slope. Then, quite suddenly but without a sound, a crack opened up in the snow above the children. In an instant an enormous slab avalanche was sliding away.

She saw Johann Jenny and Ruth Landholt thrown down by the first wave of snow. She screamed a warning to the children still on the slope but they were powerless against the surging masses. One child was able to remain upright for a few yards but then a second wave of snow overwhelmed the children. A large cloud of snow-dust developed and as it dispersed only a fearful stillness remained.

It was 12.15 hours and help was at once sent for. Of the fourteen people taken by the avalanche three were but partly covered and rescued immediately. The first outside help arrived at 13.15 though,

* Italics are the author's.

truth to tell, they were ill-equipped with rescue material. Finally, helicopters of the Swiss Air Rescue Guard flew in supplies from Zürich. Of the eleven people still missing one boy was found by an avalanche dog after three hours' burial, under 5 feet of snow. Miraculously he was still alive. But, despite 24 hours of desperate work by the rescuers, nine of the children, all aged 14, and Johann Jenny were carried away from the Porclas slope as corpses.

At an enquiry into the tragedy, it was decided that the leaders of the group were blameless—a decision which surprised a good many people. On the other hand, the grief-torn parents decided in their hearts that they were guilty and their silent reproachfulness soon caused the young schoolmistress to resign her post. In passing it should be mentioned that the standard of snow-craft demanded for a ski-instructor's certificate is not very high; it is certainly not high enough to justify much faith in its holder's ability to lead when avalanches threaten. The Lenzerheide accident rightly resulted in an outcry against improperly qualified people leading groups of school-children. Several other avalanche accidents have taken place in the company of ski-instructors and one does well to realize the limitations of the qualification once off marked runs.

On the other hand, the examination for the status of mountain guide demands a high standard of snow-craft, although some people claim that the competence of the *average* guide is lower now than it was a few generations ago. This could be true because there are many more guides than there were, and in the days when climbing had a relatively small following only the best and most resourceful men could find employ as guides. Even so, to ski-tour with a guide is usually a great deal safer than with a ski-instructor.

What is really lacking in the Alps today is a qualification as 'ski-guide'. As things stand, the guide is first and foremost a climber, not a skier; he cannot qualify without being competent on rock and ice, although how well he skis is immaterial. If, however, there were a category of ski-guides, trained in snow-craft and avalanche rescue, etc., and qualified to lead skiers on a tour, without necessarily being a rock climber, there would be many fewer parties led by ski-instructors, who are so often woefully ignorant about safety once off the ski-runs.

The main reason for relating the story of the Lenzerheide accident, however, was to make the point that those 10 young people died on a slope studded with trees. They were not the first to do so nor will they be the last, at least until the idea that a few scattered trees will prevent an avalanche is finally overcome.

The factors which help in the initial creation of an avalanche, the conception and gestation so to speak, have now been outlined as have the conditions *favourable* to a birth. But what determines and how is it determined that at a precise moment of time our Panta-gruel, our lusty and destructive giant-child should actually be born? What in fact releases an avalanche and how does the release mechanism function? Of all the aspects of avalanches it is this particular one that has captured the imagination of the most people. Those who have never been near an avalanche, and are never likely to be, assert the human voice can set one in motion. In fact, nothing about avalanches has been publicized, dramatized and occasionally misrepresented, as much as the sensitivity of their release mechan-ism. The process began as long ago as 1548 when Johannes Stumpf wrote about birds and noise setting avalanches in motion. And writ-ings about the power of the human voice in this connection later became commonplace.

Schiller wrote: '*Und willst du die schlafende Löwin nicht wecken / So wandre still durch die Strasse der Schrecken!*' (And if you don't wish to waken the sleeping avalanche then walk quietly along the road of fear.)

Byron too was captivated by the delicacy of avalanche release. He wrote: 'Ye avalanches which a breath draws down—'

There are many instances to illustrate the credence given to noise as an avalanche trigger. For example, a party crossing the Umbrail Pass in 1774, during a snow-storm, stuffed their mule-bells with cloth to silence them. They were right to fear avalanches on the Umbrail, however, because a 19th-century trader transporting salt and wine over the pass lost seven of his men and 210 horses over the span of a few years. Another example of the belief in noise-released avalanches is that of the father in the Muotatal who flew into a rage if his children slammed a door after a snow-fall.

There are also cases in which noise is actually claimed to have set an avalanche in motion. The first stroke of the church-bell at Churwalden one Sunday morning is reported to have brought two down, and an American tourist called Doctor Grant is said to have produced one by shouting at the Junction above Chamonix in 1838. In 1904, David Martin, a so-called 'observer of Alpine nature' wrote in the *Courrier des Alpes* that 'the children of Valgaudemar amuse themselves by shouting to start small avalanches which they then ride down upon'. The value of Martin's other Alpine observations is unknown to me but I trust they were more acute than this particular one!

However, before taking up a position on noise-started avalanches, the facts about avalanche release, as far as they are known, must be explained. Although it is obvious from earlier chapters that there are stresses at work in the snow cover which can produce a fracture, and from that an avalanche, we must go back to fundamentals and be more explicit.

We will begin by imagining a layer of snow on a slope of even gradient. The effect of gravity and the settling and creeping of the snow together create a state of stress within the layer. The stresses are greater or smaller according to the steepness of the slope, the weight of snow on the slope, and the viscosity of the snow. (We are imagining a slope of even gradient but it must not be forgotten that local variations in the stresses can be brought about by changes in the vertical and horizontal slope profiles, and also by variations in the thickness of the snow cover.)

The stress in the snow is counteracted by the *strength* of the snow. This is a combination of the cohesion of the snow and the static friction between the crystals. The static friction, it will be remembered, varies with the type of crystal and the amount of weight placed upon them.

Along its underside the snow layer is anchored to the snow layers below, or to the ground. Additionally, it may be anchored at the top of the slope, at the sides, and supported from below, all factors which help to hold it in place. Nevertheless, these peripheral anchorages are less important than one might expect. Their effect is only felt over a short distance because snow is a viscoplastic material; it is only in very compact snow that peripheral anchorages have a considerable influence.

On a slope of reasonable size, therefore, the factor which determines the release of a slab avalanche is the stability of the anchorage along the underside of the layer—in other words the sheer strength of the underlying snow stratum, or of the anchorage to the ground.

Following on from this statement, let us eliminate any peripheral influences, however slight, by imagining a snow layer, or an ensemble of snow layers, lying on a slope of infinite length and breadth. It will be clear that the stability, in general terms, of the snow on the slope depends on how much sheer stress there is and how much sheer strength the snow can offer to counteract it. Somewhere in the snow cover there may be a weak stratum in which the stresses imposed by the weight of snow above are bringing it close to the fracture point. Usually the overall stability of the snow depends on this weakest layer—the layer in which the margin of sheer strength over sheer stress is the smallest. Both factors can be measured and calculated in the field and are usually expressed in kilograms per tenth of a square metre.

Thinking on these lines, Edwin Bucher introduced a simple *Stability Index* in which the stability of a given stratum is the sheer strength divided by the sheer stress. If the result is greater than one the stratum is basically stable, and vice-versa if it is less than one.

It is obvious, then, that for an avalanche to occur something must happen which either increases the sheer stress, or decreases the sheer strength, until a fracture results. The process which changes the ratio of strength to stress, and releases an avalanche, may be a gradual one spread over some time; in this case it can be termed a *spontaneous release by gradual influences*. Alternatively, the strength/stress ratio may be changed by an event of some sort, a shock which momentarily causes a sudden rise in the stress forces; in this case it can be termed a *release by sudden incident*.

SPONTANEOUS AVALANCHE RELEASE BY GRADUAL
  INFLUENCES

The clearest insight into this form of avalanche release is given by seeing how certain factors can slowly cause a change in the sheer strength/sheer stress ratio, so bringing about a reduction in the Stability Index. We will begin with *factors which reduce the sheer strength of the snow*.

Destructive metamorphism in new snow. Dry loose-snow avalanches with their localized starting points have a somewhat different release mechanism to that of slab avalanches in that they begin with the loss of equilibrium of a few crystals rather than a whole stratum. However, the broad principle of stress imposed by gravity and strength created by cohesion still applies.

When new snow reaches the ground in calm conditions, the dendritic crystals interlock by means of their fine branches and spikes, as already explained. But the strength of this cohesion is soon being undermined by destructive metamorphism. The points and spikes begin to regress and the interlocking system is gradually weakened. The effect is for the static friction angle of the snow to be progressively reduced. Hour by hour the balance of the snow becomes more precarious, until the moment when the static friction angle of the snow coincides with that of the slope on which it is lying. A few crystals lose their balance at a given point; they topple, set those below them in motion, and one of those tongue-shaped, loose, dry, flowing avalanches is created. They are very common on steep slopes after a snow-fall under calm conditions, especially on slopes where the sun speeds up destructive metamorphism. Later, when the granules have started to form bonds, the snow will become stable again, but there is a definite period after a snow-fall when loose, dry, flowing avalanches take place.

Temperature rise. A rise in the temperature of the snow is one of the commonest causes of a reduction in sheer strength. It has been found that the influence of such a temperature rise is greater the more strength the snow originally had. It is also greater the nearer the temperature of the snow is brought towards freezing point. Thus, a 3° rise which raises the snow temperature to 5° below freezing will have less weakening effect than 3° rise which lifts the snow temperature to 1° below freezing. It follows that a temperature rise in spring, when the whole snow cover is relatively warm (near the melting point), produces the most marked weakening effect.

The first thing that a rise in temperature does is to weaken the bonds between the crystals; and if the rise continues it will eventually surround each crystal with a film of melt water which lubricates and reduces the static friction. The sheer strength of the snow is then reduced almost to nothing, and in this state wet snow avalanches commonly occur. The point must be made that a drop in snow temperature has the inverse effect to that of a rise.

Rain. Rain brings about a rapid rise in the snow temperature and also provides free water for lubrication.

Constructive metamorphism. The importance of cup crystals in the formation of avalanches has been frequently mentioned. If there exists in the snow cover a stratum which is undergoing constructive metamorphism, as there almost always is, then the sheer strength of the stratum is being gradually reduced. This process alone can ultimately cause the release of avalanches which occur some weeks after a snow-fall and for no apparent reason. Even if constructive metamorphism does not actually precipitate an avalanche it may well reduce the Stability Index to the point where the slightest increase in sheer stress could destroy the equilibrium.

In the other camp, *factors which gradually increase the sheer stress*, there are the following:

Snow-fall. The weight of an additional snow-fall is the commonest of all the naturally occurring influences which can increase the sheer stress in an underlying stratum until it reaches breaking point. During a snow-storm, the stability of the snow on a slope will gradually be reduced by the extra weight until the Stability Index drops to below 1 and an avalanche is released. This, of course, will occur on the steeper slopes first where a given amount of snow creates greater stress. On such slopes avalanches or small slides may recur every few hours during a storm. Herein lies the explanation for a fact which may appear paradoxical at first sight: it is on the less steep slopes that the most dangerous avalanches start because more snow will have to accumulate before the sheer strength is overcome. In addition, the lower layers of a snow-fall are always compacted and strengthened by the weight of snow above, and the effect of this is greater on the less steep slopes. Similarly, devastating avalanches have been produced by heavy snow-fall on to a fairly stable foundation. Had the foundation been less stable, the avalanche would have been released sooner when the smaller quantity of accumulated snow would have been less dangerous. In general terms, gradients of about 30–35° produce the catastrophic avalanches which occur during violent snow-storms and invade inhabited areas. The snow over a wide area of slope reaches a level of delicate equilibrium until, finally, in a single instant the whole layer is set moving. When a thick stratum of new snow is released in this way it can form a tremendous airborne-powder avalanche.

Rain. Rain has already been numbered among the factors which reduce sheer strength, but at the same time it increases stress by the addition of weight to the snow cover.

AVALANCHE RELEASE BY SUDDEN INCIDENT

It will have been noticed in the section above on Release by Gradual Influences that, in every case, the mechanism hinged on the slow reduction of the Stability Index to below 1. But in this section we shall see that an avalanche *can* occur even when the snow strata are essentially stable. A sudden increase in stress may outweigh the strength of the snow and bring the Stability Index below 1. This can be caused in two ways: by the intervention of an external agent, or by a fracture in a peripheral anchorage.

Intervention by external agent. The external agent that brings about an avalanche release can be a lump of snow, an icicle, or any other body falling from a tree, rock or cliff; a falling cornice; or a skier, climber or animal moving on to the slope.

If the intervention is of sufficient magnitude, the effect is usually immediate and violent. Occasionally, however, there can be a curious delayed reaction: several people may cross a slope in safety and then an avalanche take some of the tail-enders of a party. An interesting case of this occurred during the last war when the Swiss Army was being trained in avalanche-safety techniques at the Research Institute. Thirty men were sent across a slope which, after a recent snow-fall, was known to be avalanche-prone. They were therefore ordered to cross singly with a gap of 50 yards between each. Twenty-seven of the soldiers reached the far side in safety, and then a slab avalanche suddenly buried the last three. They were rescued within minutes and, apart from shock, were none the worse.

Skiers almost always believe that if one person has skied a slope in safety the rest of the party will also be safe, provided that they too venture on the slope singly and that there is no leviathan among them. (If there is it is usually politely suggested that he go first as the best possible test for safety.) However, cases of delayed release are surprisingly common. The probable explanation is that an underlying fragile stratum is fractured by the weight of the first person, but over too small an area to disturb the overall equilibrium of the snow cover. The weight of each successive person extends the fracture until the slab is released.

Fracture of a peripheral anchorage. So far we have ignored the effect of peripheral anchorages. We must now go back to considering them and see how they can sometimes have an important effect. Let us imagine a layer of snow lying on a slope to which it is anchored at the top, at both sides, along its underside, and supported from below. The all-important underside anchorage is stable; indeed we will assume a Stability Index of 3. Before the layer can be released as a slab avalanche, all the peripheral anchorages and the underlying one must be broken. This would appear to be a difficult condition to satisfy but, in fact, it is far from being so. *A fracture of just one of the peripheral anchorages can produce a shock sufficient to fracture the critical anchorage along the underside of the layer.* If the fractured area of the underside anchorage is then large enough to overcome the remaining peripheral anchorages, which it very often is, the avalanche will start.

One might think that, given the same Stability Index, a layer of compact snow tethered by peripheral anchorages would be more stable than a layer of loose and plastic snow in which peripheral anchorage effects are minimal. This is not so, though, because the ability of compact snow to transmit shocks can make the peripheral anchorages a liability rather than an asset. As explained above, a single fracture, termed the primary fracture, can be sufficient to bring about the fracture of the remaining anchorages. The commonest type of primary fracture to precipitate an avalanche is one caused by tensile stresses in the convex part of a slope.

However, from the number of slopes to be seen across which a crack runs, but on which the snow has not moved, it is evident that a peripheral fracture does not necessarily release an avalanche. The reason for this may be that a tensile fracture in the convex part of a slope does not bring about a fracture in the underside anchorage of the slab because the anchorage is inherently too stable. (In this

15. Slab avalanches in a gully system. A skier released the one in the centre and this caused the release of four others, two of which are visible. A rescue team can be seen sounding the avalanche snow. The dark patches in the breakaway area are conifer shrubs

16. An avalanche victim left in a grotesque position by the twisting action of the snow—an action accentuated by the skis and sticks attached to the victim's limbs

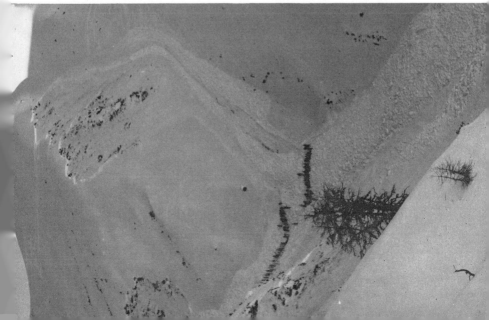

17. Melchior Schild of the Avalanche Research Institute instructs a course in the use of sounding rods

18. No exercise this time. A young German skier lies buried in an avalanche he released when skiing this steep slope. Despite a rapid rescue his life was not saved

connection, the famous mountaineer and avalanche scientist, André Roch, who retired from the Institute a few years ago, made an interesting discovery: from measurements taken at the fracture of slab avalanches released by a peripheral fracture, he has reached the empirical conclusion that such a fracture can provoke an avalanche when the Stability Index is as high as 4.) And a short tensile fracture in the convex part of a slope can fail to release an avalanche because, even if the underside anchorage of the slab breaks, the lateral anchorages just *may* hold, especially if the slab is compact.

Fractures in compact snow can travel large distances, on occasion many hundreds of yards. It is for this reason that a skier can sometimes be responsible for the remote release of a slab avalanche. He can walk along near a slope when suddenly, with a loud and frightening *Wummmph*, a fragile stratum in the snow cover, usually one of cup crystals, breaks down under his weight. There is a shock beneath his skis and one or more cracks may streak away across the snow. These progressive fractures travel at up to the speed of sound and follow the critical stress lines in the snow. Frequently, one will run up into a slope where it becomes the primary rupture which breaks the slab's other anchorages, and an avalanche is thereby released.

There remains another basic difference between a Spontaneous Avalanche Release by Gradual Influences and a Release by Sudden Incident. In the first case, when the Stability Index is in the region of 1 over a wide area of slope, the propagation of the initial movement caused by the fractures presents no problem. In one fell swoop the whole stratum will collapse. Once this has happened, the only restraining influence on the avalanche is the small one of the kinetic friction for the type of crystals involved. Assuming that the angle of the slope is greater than the kinetic friction angle of the crystals, an avalanche is bound to take place.

On the other hand, when the Stability Index is high, as we have seen it can be for a Release by Sudden Incident, the initial movement caused by the fractures does not always lead to an avalanche. Even when a slab near the top of a slope has been freed by all the necessary sheer and tensile fractures, it may continue to be supported by the still-anchored snow lower on the slope. For the initial movement to be propagated, and an avalanche occur, the effect of this support must be overcome. This can be brought about in two ways: either the weight of snow initially liberated is great enough to force into motion the snow lower on the slope or, as is quite common in the case of slab avalanches, the released snow slides up on to the surface of the snow cover. The point at which the released slab slides out on to the snow cover is called the *Stauchwall* (see Fig. 11). This

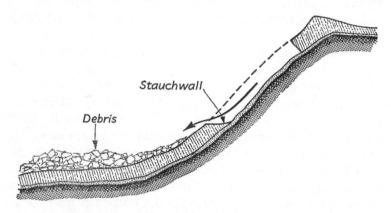

Fig. 11. Cross-section of a slab avalanche in which the released snow has slid up on to the surface of the snow cover forming a *Stauchwall* or 'buffer line'

German word is impossible to translate exactly though in America the terms 'pressure wall' and 'buffer line' have been used. It should be added that not always does a slab avalanche need to propagate its movement by setting the snow lower on the slope in motion, or by forming a *Stauchwall*. It can occur that the complete snow layer, from top to bottom of the slope, is released as the result of a sudden incident. This is especially so when the slab is compact and the stability is low, because fractures can then travel considerable distances.

The release mechanism of avalanches has now been outlined and it may appear that, given the necessary data about the snow cover, one could predict its future behaviour and the moment of release. But in this chapter, for the sake of simplicity, many subtleties have been left out. Many of the factors which have been placed in either the *stress increasing*, or the *strength decreasing* category belong to a certain extent in the other as well. We have touched on this when stating that a fall of new snow increases the stress by its weight, but also adds to the strength of the lower strata by compacting them and increasing their cohesion. And there are several factors with similar double effects. For example, melt water, which reduces strength by breaking the bonds between the crystals and lubricating them, also, up to a certain point, increases cohesion by capillary action. Cup crystals, which weaken the snow, have a higher kinetic friction angle than the crystals from which they grow. A rise in temperature, which weakens snow, also increases its plasticity, and so reduces its shock-transmitting ability. It can be said that almost every avalanche-provoking or avalanche-preventing factor has some inverse effect and so the interplay is, to say the least, complex.

It is interesting once more to consider sound as a possible avalanche trigger. There is every reason to suppose that, if the Stability Index of a given layer of snow was hovering above 1, the slightest tremor could trip the balance. The possibility of sound waves doing this cannot be denied, especially those resulting from heavy detonations. Certainly a sonic boom could release avalanches, if its window-breaking capabilities are anything to go by; even the engine noise from a low-flying jet, for example an aircraft climbing out of a valley on full throttle, could perhaps be sufficient to do so.

It seems that the most rational view to take over less powerful sound waves releasing avalanches is that, where the snow is teetering and perhaps about to avalanche anyway, a small noise could just conceivably precipitate the release. Logically, the human voice can not be excluded from the noises which might bring this about, but I am convinced that 99% of those who claim to have released avalanches with their stentorian voices have in fact done so with their clod-hopping boots. As for David Martin's observation about the children of Valgaudemar, it seems obvious that if they were close enough to ride down on the avalanches then their feet released them rather than their voices.

There remains just one intriguing aspect of avalanche release to examine. Many eminent British skiers, and especially the late Sir Arnold Lunn, have written that in spring the shadows of evening striking a sun-warmed slope caused the release of wet-snow avalanches. Sir Arnold, in his classic book *Skiing at all Heights and Seasons*, said that he had noticed the phenomenon frequently; and on one ski-descent he even raced against the advancing shadows which were releasing avalanche after avalanche.

A seemingly perfect explanation was put forward. During the day the warmth of the sun produced melt water which filled the pores between the snow crystals. When the shadows hit the slope the melt water froze almost immediately, and the 11% expansion of water when it freezes was sufficient to destroy the equilibrium and cause an avalanche. The accuracy of this once wholly plausible explanation in now in doubt however. An English glaciologist, J. Hollin, made some tests in the early 1960s and produced an interesting paper called: 'When wet snow freezes.' For his experiments he was forced to distinguish between two types of wet snow: snow whose pores were completely filled with melt water, which he termed *wet wet*, and snow in which each crystal was merely covered by a film of water, which he termed *dry wet*. Because melt water percolates downwards as it is created, he never found, in nature, the *wet wet* type in the top inches of the snow cover. Owing to the fact that a shadow passing over a snow slope initially causes only a very shallow surface layer to freeze, Hollin carried out his main freezing tests with *dry wet* snow samples. He discovered that *dry wet* snow *contracts* as it freezes, the contraction fitting the thermal coefficient of ice. However, ice did bulge from unconfined samples of *wet wet* snow when they were frozen. For his initial tests Hollin reproduced the conditions found in nature when a wet-snow cover is subjected to freezing for some time. He filled plexiglass cylinders with snow and allowed it to thaw until there was *wet wet* snow at the bottom of the cylinder and *dry wet* at the top. He then refroze the snow. Nowhere was the cylinder expanded. The *dry wet* snow shrank away from the cylinder walls and the expansion of the *wet wet* must have taken place upwards into the *dry wet*.

Expansion in wet snow as it freezes can, therefore, be ruled out as the cause of 'evening avalanches'. But why should not the contraction which Hollin found likewise increase the stress and release an avalanche? It is certainly possible but, in the first place, statistical evidence of 'evening avalanches' as a phenomenon is still lacking. Some precise observations would be interesting. For the moment, the avalanche-releasing effect of shadow striking sun-warmed snow must remain one of the mysteries which make avalanche release so difficult to pin down.

7
Death and Survival

Anyone who falls into the merciless clutches of an avalanche is subjected to one of the most terrifying experiences that man can undergo. Some victims, perhaps fortunately, die within seconds; others live for a few minutes, and a few last for several hours. The duration of survival is determined by many things, not least among them the moral and physical strength of the victim. Even when other factors are favourable it requires great stamina to live for several hours in an avalanche; and very rarely does a man show the tenacity necessary to endure a complete day. Yet in 1951 there occurred such a prolonged fight against death in an avalanche that one can but wonder at the existence of such an indomitable spirit. A young Austrian called Gerhard Freissegger survived the torments of body and mind imposed by a burial lasting $12\frac{1}{2}$ days; his survival is numbered among the great feats of endurance of all time. Today he is a living monument to the pinnacle of courage and will-power that some men can attain in time of need: not the hot-blooded flash of valour in the battle-field, but the sustained courage of a drawn-out fight for existence against all odds—a fight grimly fought, alone and in the dark. The facts of Gerhard Freissegger's ordeal speak for themselves.

Near Heiligenblut, Austria, in the early 1950s, a dam for a large hydro-electric scheme was being built on the so-called Margaritze. A two-stage cable-car had been constructed for the transport of materials and 26-year-old Gerhard Freissegger was employed at the middle station on the Sattelalp. Normally three men worked there, but January 20th, 1951, was a Saturday and, although it was really Gerhard Freissegger's free weekend, he had offered to stay so that a married colleague could go home to his wife. He smiles now as he remarks that if he had had any notion of what he was letting himself in for, he would not have been so generous.

On that January Saturday it was snowing hard, and it had already been snowing for several days. More than 3 feet of snow had fallen and during the course of the afternoon Freissegger noticed that the wind was getting up. By 18.00 hours, when Freissegger and his colleague, Siegfried Lindner, were ready to close down the station, the storm had intensified. Freissegger set out uphill to the living hut some 50 yards above the station and Lindner, who was talking on the telephone, followed a few minutes later.

The three-roomed hut was cold and they cooked and ate their meal in near silence. Freissegger is, in any case, taciturn and withdrawn and a few comments on the extraordinary snow-fall and the hope that it would soon stop, made up the conversation of that evening. They were about to go to bed when Hartiger, one of the two men in the top station, the Winkelstation, rang up to say that they had been warned of avalanche danger. It was believed that the Winkelstation was more endangered by avalanches than the station on the Sattelalp, and Lindner jokingly said over the telephone: 'Don't worry. Gerhard and I'll come up and dig you out in the morning if anything happens.'

The room where Freissegger usually slept was so cold that he decided to sleep in the main room, in the bunk above Lindner's, vacated by the colleague who had gone home. They undressed to shirt and underpants and by 19.30 they were both lying down. The storm was howling ever louder and about midnight it was raging so violently that Freissegger awoke. He lay for a while listening to it and the thought occurred to him that the roof might blow off. Then he dozed once more.

Sometime about 02.00 he was woken again, this time by Lindner's voice: 'Just listen to that wind, Gerhard.'

But Freissegger had no time to reply. There was a splintering crash and he was struck by a blast of snow. Instinctively he threw up his arm to guard his face, and the next moment he was clamped down solidly on his straw mattress. His legs were completely immobilized and his left arm, which had been under his pillow, was also held firmly. He was covered in snow apart from a small space over his face and chest which had been kept clear by a roof timber. With his right hand, the only part of his body he could move to any extent, he brushed the snow from his nose and mouth and found that he could breathe freely.

The he heard Siegfriend Lindner once more. 'Help me, Gerhard! Help me!' he groaned.

'I can't move myself or I would. We'll have to wait for help,' replied Freissegger.

Lindner's entreaties continued for what Freissegger believes must have been four hours, and all the time Freissegger tried to comfort him with assurances that rescuers would soon arrive. But Siegfried Lindner's voice became weaker and weaker, and then his groans were less frequent. Finally, there was no reply when Freissegger called to him, and nor was there any sound of breathing. In that moment, the realization that he was alone came to Freissegger with shattering impact.

Of those first few hours Freissegger has little memory, other than that his feet and legs were going numb and that he never doubted a speedy rescue. And, sure enough, the heavy silence was suddenly broken. The rescuers had arrived. Freissegger could hear them walking about above him; he could hear their voices and the sound of their shovelling and probing as they searched for him and Lindner. There was gladness in his heart, for surely they would soon find him. He called and shouted. During all that long day the sounds came and went above him, his hope coming and going with them. And then the sounds became ever more distant and, finally, that heavy silence returned to prey upon his mind.

After what seemed like weeks the voices were there again, and during that second day's search a sounding-rod actually hit a piece of timber right next to Freissegger, so close to him that he felt the vibration of its impact. He shouted again and again without making himself heard. He was suffering the same bitter experience as so many others before him: while an avalanche victim can usually hear noises from the surface, his own desperate cries seldom reach beyond the snow enclosing him. Help is but a few feet away, clearly audible, yet all the attempts of the victim to make his presence known are in vain. The anguish of the situation could unhinge a man's mind— but not that of Gerhard Freissegger, even though the rescuers were close by for three days, and he was to hear other people regularly throughout his burial.

By the fourth day he was sodden with melting snow, but his left arm was free, and so were his legs down to the knees. He fought against the temptation to sleep because he believed he would not wake again. He talked to himself interminably, and he sang to try to keep his spirits up. He cleared the snow from around his body to give himself more movement, and then he began to scratch methodically with his finger nails at the hard snow above him.

And when the fear of death pressed upon him most insistently, he thought of all the people he wanted to see again and of his plans for the future. That flicker of life in the midst of the snow was kept burning by the unfaltering power of his mind to fight off despair.

Twice a day for several days after the rescuers had left, a group of Freissegger's colleagues carrying materials and equipment passed within 20 yards of where he lay. Each time he gathered all his strength to shout and each time, as their voices faded into the distance, he overcame the temptation to give up hope.

His progress in digging upwards was hampered by his trapped legs which limited his range of movement. Finally, on what he estimates was the eighth day, he managed to free them, swollen, insensible and useless limbs. A little later he came across a splinter of wood, probably off his bunk, and with this he continued to bore upwards. He could only work a little at a time, falling back in exhaustion after each short bout. He ate snow in large quantities to assuage the burning thirst that possessed him, and he forced himself to continue digging. The snow he was dislodging fell on him and melted, but he was beyond the realms of feeling. When he could no longer reach the roof of his tomb he dragged himself into a kneeling position and struggled on. He was still fighting the temptation to sleep, but by now he was dozing more and more.

Then he noticed that there was a faint patch of light in the blackness above him. It was probably the morning of the tenth day. He worked desperately on, but the limits set by his exhaustion were so stringent that it was several hours before his piece of wood pierced the snow and cold air rushed in. With renewed hope, he widened the hole and forced himself upwards until he could see out. Dusk was falling and the bleak landscape was deserted. He shouted once, feebly, and then sank back into his hole. He could have crawled out, but it is extraordinary that, even at this stage, he was lucid enough to realize that he would freeze to death outside in his shirt and underpants. The temptation to escape the tomb overcome, Freissegger settled back to await the group of porters, even though they had not passed by for several days.

His fitful doze was interrupted by a noise. Instantly he cried out and began to struggle towards the opening above him. As he did so a flurry of snow fell down it; and when he looked out he saw two skiers disappearing. They had passed so close to his hole that snow had been swished into it by their skis.

For the first time he gave in to the belief that he would die. He slumped back into the hole, utterly defeated. But gradually his will to live returned, although by now his body had all but reached the end of its tether. He did not know it, but it was the evening of the eleventh day.

He dozed again, and this time the sound of the approaching group of porters woke him. He put all the remaining vestiges of his strength into a shout. And they heard him. Looking around them for the origin of the strange and muffled sound they saw a hand protruding from the snow. It was midday of February 2nd. With astonishment and joy they freed the colleague they had long thought dead and carefully took him to the cable-car. In a weak but clear voice Freissegger told them about his ordeal and, with his sardonic humour, remarked that he would like to eat a goulash. When asked for how long he thought he had been buried he replied that it must have been two months.

After being strapped to a ladder and taken down to the valley he was transported to Lienz hospital. There he was lucky to come under the care of Professor Bernard who had been doctor to the German Nanga Parbat expedition and was well qualified to deal with Freiessegger's ailments. He had third degree frost-bite of both feet, inflammation of the bladder, lungs and kidneys and spasmodic disturbances in his vision. He had lost 65 lb in weight. To the doctor's complete astonishment, however, there were no discernible psychological effects. But his mother says that some idea of the mental torture he had undergone is given by the fact that he had clawed his thighs until they were raw.

For two days he lay in hospital with the lower part of his legs exposed. They were smooth and deathly white, as if of alabaster. And then the toes slowly began to blacken. Shortly afterwards the first of a series of amputations took place—a series which only terminated when he had lost both legs below the knee. Never once did Freissegger complain, but he gave an insight to his feelings before the first operation. He and his mother were contemplating his blackening toes when he said: 'Mother, am I going to lose both feet?'

'I don't know,' she replied, 'perhaps just one, Gerhard.'

'Oh God, I hope it's only one,' he said.

After a six-week period in Lienz hospital, he went to a nerve hospital in Klagenfurt for eleven weeks. There they cured a complaint which had prevented full use of his left hand, and he received artificial legs. By the autumn he was back at work with his firm, this time as telephonist on the Margaritze dam project. Despite press reports at the time, snow and mountains held no terror for him; he spent that winter at the Winkelstation, the station just above the scene of his ordeal.

Now, Freissegger is a storeman at the head office of his firm. He suffers much pain and further amputations are being considered. He appears to view the prospect with equanimity, but what goes on inside that exterior of iron? One might think that he is impervious to suffering from the modest way he speaks, but the odd remark passed by his charming wife leaves no doubt that he suffers as much as anyone. He, however, will not talk about his feelings or reactions, and, after a single press interview following the accident, he resolutely turned down all offers from press, radio and television. I was therefore fortunate that he agreed to see me and help me with his story; and, this apart, it was a privilege to come face to face with such a remarkable man.

*   *   *

Victims of an avalanche may die from a variety of causes; and they may die either quickly or slowly. In the first place, victims may be flung against obstacles in the avalanche path, or there may be rocks, trees and ice blocks in the avalanche itself. Such dangerous fellow travellers can inflict the most hideous wounds. For example, Johann Coaz quotes the case of three farmers who went to fetch hay from a hut. On their way they were surprised by an avalanche but two of them were able to free themselves unharmed. The third man had had his legs so torn apart that his entrails were showing and he died within minutes of being dug free. In fact, Coaz does not state whether the injury was caused by obstacles in the avalanche path, debris in the avalanche, or by the twisting and wrenching effect of the moving snow alone. This latter can, as we have already seen in the case of Zdarsky with his 80 dislocations and fractures, cause grave injuries.

And if a victim is caught with his skis on his feet and with the loops of his ski-sticks passed over his wrists the twisting action of the snow is particularly dangerous. The skis and sticks form extensions of his limbs and gives the snow more leverage. A glance at photograph 16 illustrates this. Note, too, that the victim's hands which are showing between his legs, have been pulled down by his sticks. The lethal side effect of this is to force the victim's face forward into the snow so that he is bound to suffocate.

If an avalanche victim loses consciousness, either owing to a blow on the head or to increased adrenaline output brought about by fear, his survival chances are much reduced. If he is unable to help himself at all, a thin covering of snow, or even a mouth and nose full, could well suffocate him. In fact, loss of consciousness caused by fear, the swoon of the romantic novel, is usually reversed once the person is lying horizontally, and the blood supply to the brain is restored. It is unlikely, therefore, that fear alone could produce prolonged unconsciousness in an avalanche.

Suffocation, or hypoxia (lack of oxygen), is the cause of death that most people associate with avalanches, but there are three distinct ways in which an avalanche can suffocate its victims: two of them cause rapid death, and the other slow death. Firstly, the weight of snow on the throat and chest can make it impossible to draw breath. It may seem hard to credit, except in cases of very deep burial, that the pressure of snow on the body can be so great that one cannot even make the small movement necessary to fill one's lungs. It depends, of course, on the density of the snow but I have heard personally from a man, who was buried only 3 feet deep, that it was like being encased in concrete. He had lost a glove in the avalanche and the grip of the snow on his fingers was so vice-like that he could not flex the joints, even fractionally. He was fortunate to be freed very quickly by friends.

On the other hand, avalanche snow can, on rare occasions, be quite loose. In 1953 there was the truly remarkable case of a forestry worker who was buried, in a vertical position, in loose and porous snow. He managed to free his hands and, by working with these, and by wriggling his body and legs, he gained freedom of movement. Then, systematically, he pulled snow from above his head, passed it down his body and stood on it—thereby raising himself slowly in the avalanche. After one and a half hours of excruciating effort, he broke surface and hauled himself out of the hole, utterly spent.

But to return to suffocation and hypoxia: the second cause of this in an avalanche is the inhalation of snow. A natural reaction of many people caught by an avalanche is to gasp with fright, and in doing so they may inhale quite large quantities of snow. During the descent of the avalanche, more snow may be forced into their mouth and nose until the air passages, including the trachea and bronchii, are blocked. Suffocation is almost immediate.

The third cause of suffocation is the gradual exhaustion of the oxygen supplies around the victim as he breathes in the confines of his position in the snow. This suffocation may be spread over some time and depends to a large extent on the porosity and air content of the snow. Anything which reduces the oxygen requirements of the body will prolong the victim's life and give more time for a successful rescue. Loss of consciousness is beneficial in this case because it reduces the oxygen demands of the tissues, and particularly of the brain whose need is the greatest. Cooling of the body by the surrounding snow is also a factor in reducing oxygen requirements; indeed, it is for this very purpose that hypothermia (deep-freeze surgery) is practised.

It has been discovered during hypothermia that at a body temperature of 92°F, oxygen consumption is 80% of normal; at 86°F it is 70% of normal; and at 74°F it is only 50% of normal. Shivering increases consumption, and during an operation relaxants are given to prevent its occurrence. Below 74°F the hypothermia itself causes unconsciousness and if the temperature is reduced too far ventricular fibrillation sets in with, in dogs, spontaneous arrest of the heart at about 60°F.

If an avalanche victim has not died from one of the causes so far mentioned, there still remain frost-bite and exhaustion to finish him off, as they almost did Gerhard Freissegger. When someone dies after a long burial, the extent of the part played by these factors or by hypoxia is not known.

Mountain people talk of one other form of death in an avalanche— 'shock death'. When a victim is dug out of an avalanche and he shows no sign of having suffocated, that is to say that his face is not purple but is grey or white instead, the people say that he died of shock. And this view is reinforced if there is no sign that the victim lived long enough to try and free himself. Basically, the belief is that the victim instantaneously died of fright. A large number of avalanche victims, perhaps a third, are said to have died in this way.

A famous Zürich anaesthetist, Dr. Hossli, does not accept this as possible and for some time he has been trying to find another explanation for the death of victims who, even when freed from an avalanche very quickly, cannot be revived by artificial respiration and show no outward signs of having suffocated. Dr. Hossli points out that if these avalanche victims had really died of fright alone, then many people would die for the same reason in other dangerous situations. For example, people would drop dead when narrowly missed by a vehicle in the streets, and so on.

Dr. Hossli believes that the real reason for so-called 'shock death' is vagal effect. The vagus nerve is mainly responsible for controlling heartbeat, and there are certain parts of the body in which it can be irritated by an outside influence. One of these irritation centres is the solar plexus and another is the epiglottis area, or back of the throat. An irritation to the vagus nerve in either of these areas produces certain reactions in the body, such as bradycardia (slowing of the heartbeat), a spasm of the larynx and vomiting. Classic example of vagal effect are those resulting from a punch in the solar plexus, and food or drink 'going down the wrong way'.

It has already been explained that many people inhale quite large quantities of snow when caught in an avalanche, and Dr. Hossli thinks that even small quantities could produce such a strong irritation in the epiglottis area that the resultant vagal effect would be very strong also. When buried in snow the effects, particularly of the laryngal spasm, could be fatal. The victim would quickly die of hypoxia. And quite often victims are found to have vomited, which reinforces Dr. Hossli's theory of vagal effect; and this too could be fatal because the victim might inhale the vomit.

Dr. Hossli does not think that someone who had died of hypoxia as quickly as this would necessarily be purple in the face, particularly bearing in mind the reduced blood circulation brought about by bradycardia. A victim freed from an avalanche in the exceptionally short time of 10–15 minutes could be past revival and grey faced, and the rescuers could easily conclude that he had died instantaneously, of some cause other than hypoxia.

If the Hossli theory is correct, as many doctors seem to think it is, then it is particarly important not to inhale powder snow when caught in an avalanche. Even without vagal effect, there is a chance of blocking the air passages to the lungs. It is therefore a basic rule to keep one's mouth clamped shut in the moments of being carried away by an avalanche.

Avalanche victims who are buried in a building have a much better chance of survival than skiers and others who are completely enveloped in snow. The case of Gerhard Freissegger was really neither one thing nor the other because, although much of his body was in contact with snow, the space above his face and chest contributed much to his survival.

In passing it must be stated that Freissegger's 12½-day burial was not the record—if one can use such a term. In 1775 three Italian women stayed alive for 37 days under an avalanche, but their bodies were not in contact with snow. The facts of the case are written in a book of 1765 with the pleasing title: *A True and Particular Account of the most Surprising Preservation and Happy Deliverance of Three Women buried 37 Days by a heavy Fall of Snow from the Mountains at the Village of Bergemoletto in Italy.* The author, Ignazio Somis, was a professor of medicine at Turin and he attended the women after their rescue.

The avalanche struck the village of Bergemoletto in the Stura Valley, Piedmont, just before Mass on March 19th, 1775. A 45-year-old woman, Maria Anna Rocchia, her sister-in-law Anna Rocchia, 13-year-old daughter Margareta and 6-year-old son Antonio, were trapped in their stable with all the livestock. The roof had caved in leaving them a space 12 feet long, 8 feet wide and 5 feet high. During the first days all the livestock died apart from two goats, one in full milk and the other due to kid in mid-April. Anna had fifteen chestnuts in her pocket and when these had been eaten there remained just a quart of milk a day on which to subsist.

Antonio died on about the twelfth day after suffering violent stomach pains. By then, however, the others were hardly hungry. In the very early days the passage of time had been marked, or so they believed, by the crowing of the cock trapped with them, but after its death their only indication of time was when the goat produced its kid in mid-April. They killed the kid at once so as to have more milk for themselves.

April was a warm month, and on the 24th the men of the family, who had not been buried, resumed their search for bodies. Their rescue attempts immediately after the avalanche had been given up as hopeless because the snow was some 60 feet deep. On the 24th, however, they finally located the stable and were astonished to find the two women and the girl still alive. Strange to relate, Maria Anna's brother had previously had a dream in which his sister was still alive and he had spoken to her.

None of the victims could walk when rescued, though the two younger ones, Anna and Margareta, soon recovered. Maria Anna was far worse affected. She was bald, saw double, could not sleep and was only able to walk again after many weeks. She had headaches and 'the pupils of her eyes trembled well into July'. The victims related that the greatest hardship of their ordeal was the stench of putrefying corpses and the constant dripping of snow-water.

Their clothes were, in fact, rotten and the account says: 'Maria Anna's shift was little better than lint and so impregnated with filth and nastiness that four washings and a boiling in lime were hardly sufficient to clean it again.' (They must have been thrifty people not to have burnt it on the spot.)

It is interesting to note, incidentally, that throughout the English translation of Ignazio Somis' book the original Italian word for avalanche, *valanca*, was retained because there was no English equivalent at the time.

*   *   *

Among avalanche victims like skiers, whose bodies are in total contact with snow, successful rescues are fewer and invariably after a much shorter burial time. Statistics over the years show clearly that the overall survival chances for a skier caught by an avalanche are just under one in three. It is estimated that approximately 20% of those taken in the open by an avalanche die almost immediately from one of the rapid types of suffocation or from traumatic injury. For the remaining 80% the depth and duration of their burial are crucial in determining their chances of survival.

From avalanche accident and rescue reports of the last 20 years or so, it has been possible to draw up a fairly precise picture of the *average* survival chances as a function of the length of time buried. As mentioned in the foregoing paragraph, 80% of those taken by an avalanche survive the immediate effects, therefore if dug out *immediately* there is an 80% chance of their being alive. After one hour, that chance has gone down to 40%; after two hours it is down to 20%; after three hours it is 10%. From three hours onwards, the chances continue to dwindle until theoretically they reach zero after seven hours of burial. With regard to depth of burial, it is very rare to find anyone alive if buried deeper than about 2 metres. Therefore, the survival chances expressed in percentage above are bound to reflect the situation for those buried at shallower depths, as is usual anyway.

The conclusion to be drawn from the above statistics are: *if an avalanche victim caught in the open is not found within two hours, or if he is buried at more than 2 metres' depth, there is less than a 1 in 5 chance of rescuing him alive.*

As one would expect, there have been exceptions to this rule and hope for the survival of an avalanche victim should, of course, not be abandoned after two hours. There is, for instance, the famous story of 'Avalanche Franz-Joseph', a courier who used to drive goods and people along the Flexenstrasse between Stuben, Langen and Lech, Austria, in the 1880s.

On December 21st, 1886, Franz-Joseph set out from Stuben in a snow-storm with a load of flour. About mid-morning the storm had become so intense that he decided to turn back, only to find the road blocked by an avalanche which had come down since he passed by. He had set to work with a shovel to clear a way when a second avalanche struck and carried him almost 1,000 feet down from the road.

He was buried, but only under a light covering of snow. He had a leg broken in two places and a gash on the head, but with great energy he began to free himself. He had partly succeeded and there was a free space around most of his body when yet another avalanche came down, burying him deeply. He was left in total darkness, facing uphill in a semi-sitting position. He beat at the surrounding snow, but the air soon became hot and heavy and he lost consciousness.

He came to his senses again with a cold feeling around the legs and the air seemed fresher. He soon realized that he was in a stream-bed and that the water had thawed its way through to him, bringing air with it. Undoubtedly the stream saved his life—but it almost

a. Melchior Schild sets his avalanche dog, Iso, to work
b. Dog and master dig together

19c. Burkhard Beusch of the Parsenndienst carries out artificial respiration with 'AMBU' equipment

19d. The revived victim receives tea and a hot water bottle
These photographs were taken during a practice but the girl was indeed buried and Iso located her

drowned him first. In his tomb the water level rose steadily. It climbed up his body until he was craning upwards to keep his head clear. It was at his chin, and he was almost demented with fear, when it thawed its way out of the chamber again and the level dropped. It left him numb with cold and weakened, but at least he had air.

Shortly after the avalanche had swept Franz-Joseph away, the local policeman had happened to pass along the road. Finding the master-less horse, shivering and with icicles hanging from its belly, he and the postman, who had also arrived, soon drew the correct conclusion and went for help. The short afternoon and the snow-storm prevented any rescue attempt that day, but the next morning was fine and there were soon 40 men at work on the avalanche debris.

Franz-Joseph was woken by voices and the noise made by the two-man sounding-rod normally used for locating the road after a heavy snow-fall. He knew at once that the rescuers must think him dead because such a heavy rod could cause terrible injury to a live person. He hardly knew whether to hope they would find him with it or not.

For several hours the sounds of the search came and went, and Franz-Joseph dozed; but he was woken suddenly by the noise of the rod near him. In an agony of suspense he waited to see where it would strike next time. He heard the hiss as it slid down through the snow towards him, and it grazed his shoulder before thudding into the stream-bed below him. He shot his hand across and grasped it firmly. On the surface, those wielding the rod shouted that they had found him, still alive. The others showed disbelief until the rod was seen unmistakably to move. The rescuers rushed to dig the victim out and after a burial of 29 hours 'Avalanche Franz-Joseph', as he was thereafter known, was carried down to Stuben where the women had already lit candles for his soul.

The facts of this story cannot, unfortunately, be verified, but there is little reason to doubt its main features if one considers the fully authenticated case of a young woman who survived a burial of 21 hours in far less favourable circumstances. She was completely surrounded by snow and there were no factors like an air passage cut by stream water to help maintain life. Her case is one of the longest known survivals of a skier buried in an avalanche in full contact with the snow.

In the afternoon of Sunday, January 21st, 1945, a 21-year-old farmer's daughter set out from her home in the tiny village of Martisberg, Valais, to visit her fiancé in a neighbouring village. The weather was very bad however, and after a while she decided to turn back—only to be buried almost immediately in a small avalanche. There was no alarm, either in her own home or in her fiancé's home, for in both places it was assumed that she had stayed in the other because of the storm. There was no telephone link between them to check on her safety.

Only on the following morning was it discovered that she was missing. The search soon concentrated on the debris of a small avalanche in a gully her route had to cross. Naturally no one hoped to find her alive after a burial in excess of 20 hours. The rescuers were therefore staggered when they finally found her and the 'corpse' opened its eyes and spoke to them. She was stiff and a doctor discovered slight frost-bite of feet and hands. She recovered fully, however, married and became the mother of a large family. Her survival was doubtless due to several things: the light coma she fell into, her warm clothes, and the fact that the avalanche snow was shallow. This latter meant that she was near enough to the surface to have air, yet near enough also to the ground to obtain some degree of warmth.

An extraordinary survival took place when an avalanche destroyed the major part of a mining camp at Granduc, British Columbia, in February, 1965. A 32-year-old carpenter was in the open when the avalanche came down. It came silently and he was unaware of it until it struck him from behind; when he came to his senses again he was buried in snow, lying on his side with his right arm clamped beneath him. There was a space around his head and he could move his left hand.

He lost and regained consciousness periodically and he had no idea of the passage of time. He could hear nothing, even though there was a helicopter landing and taking off from an emergency pad overhead. He did not know, of course, that most of the camp had been destroyed and that some 25 other men were missing. He therefore thought that the rescue operation would be concentrated on looking for him; and he expected, during his periods of consciousness, to be dug free at any moment. This prevented him from losing hope.

Almost 79 hours after the avalanche, a bulldozer was clearing snow when the blade exposed something that looked like a bundle of clothing. Two members of the Vancouver Mountain Rescue Group jumped down from the blade, where they had been riding, and began to dig. They had cleared the snow from the upper part of a man's body when suddenly he blinked and said distinctly: 'Watch my legs!' The astonished rescuers lifted him carefully from the compacted snow and placed him on a stretcher. There was great excitement because any hope of locating further survivors had been abandoned many hours before.

There was no doctor at the camp when the carpenter was found but he was soon in a helicopter on his way to the hospital at Ketchikan, Alaska. He had lapsed into unconsciousness on the stretcher, but he came to again in the helicopter and was able to nod when asked if he was all right. At Ketchikan hospital he came under the care of Dr. James Wilson and his team.

According to Dr. Wilson the patient was in a very grave condition on arrival. He had severe frost-bite of the extremities, particularly of the left foot and right hand, and there was a marked electrolyte imbalance of the blood brought about by hypoxia, dehydration and partial starvation. But the most advanced techniques of medicine were used, and the patient responded well to them. He underwent a continual process of monitoring and testing as doctors checked his blood chemistry and corrected it as necessary by adding chemicals intravenously.

Hyperbaric oxygen treatment was also used. This consists of placing the patient in a chamber and pumping it full of pure oxygen at up to three times atmospheric pressure. In this way it is possible to increase the oxygen supply to the body tissues by as much as 18 times. The technique had been used with success in cases of gas gangrene but, at the time, it had not been tried in the treatment of frost-bite. It worked very well and the carpenter's subsequent amputations were certainly less than if it had not been used.

In the last five years there have been two or three other cases of people surviving between 20 and 30 hours in full contact with the snow. But such events are so extraordinarily rare that they have little noticeable effect on the statistically-derived curve of survival chances, measured against burial time, outlined earlier.

*　　*　　*

Such remarkable survivals as have been quoted above in no way alter the fact that to be in avalanche is to have at least one foot in the grave, and more usually both. It follows that prevention is better than cure, and anyone who wittingly places himself in avalanche danger is a fool. Nevertheless, occasions can arise when, with the best will in the world, one cannot avoid the danger; this may be so if the weather deteriorates while one is on a ski-tour.

It is known that about 90% of avalanches which kill skiers are released by the skiers themselves, and usually they are taken completely by surprise. This surprise element gives the avalanche a much better chance of killing its victims; therefore, one of the first assets for survival when menaced by avalanches is to recognize that the danger exists. This calls for a knowledge of the facts of avalanche creation, explained earlier, coupled with specific information about the conditions being encountered. Among other things, one must know what sort of weather has prevailed in recent weeks and in what direction any strong winds have blown. One must then be able to vizualize where those winds will have deposited large amounts of snow and choose a route to avoid such places.

The properties of the underlying snow strata must also be known and, if in doubt, it may well be worthwhile to dig a hole and have a look. Alternatively, an approximate idea can be obtained by pushing an inverted ski-stick into the snow cover and feeling whether compact strata are lying on looser ones lower down. There is the quaint story of a well-known guide called Maurice Crettex, of Champex, who used to push his ice-axe into the snow, put his ear to it and tell clients he was listening for avalanche danger. Of course he heard nothing, but pushing his ice-axe into the snow had already told him what he wanted to know.

With regard to terrain, we have already examined such factors as concave and convex slopes, slopes with rocks or cornices at the top, woodland and so on. It goes without saying that in times of danger a route must be chosen which remains in thickly timbered areas and along ridges wherever possible.

It is unwise to cross a slope at a point from which the top or bottom of the slope is invisible, unless one already knows what lies out of sight. Without knowledge of the type of terrain above, one cannot accurately assess the chance of an avalanche occurring. In addition, the knowledge that lower down a slope there are rock outcrops and vertical drops, over which even the smallest avalanche could prove fatal, might well influence a decision whether to cross the slope at all.

Indeed, when choosing a route one must never forget that, despite the best of caution, one may still release an avalanche. The likely consequences of this must be examined before it is given a chance to occur, and one must then balance in one's mind the avalanche probability against the avalanche consequences. For example, it may be preferable to cross a slope on which an avalanche is quite likely, but on which it would only run a short way over smooth ground, rather than cross a slope on which an avalanche is less likely but on which a victim would be carried a great distance—and perhaps bounced over a few rock outcrops for good measure. Similarly, although there is more chance of starting an avalanche by crossing a slope in its convex part, it may be better to do so and be on top of the avalanche when it breaks, rather than cross lower on the slope where there is less tension in the snow cover, but where an avalanche from above would bury one deeply.

Enough has already been written about steepness in earlier chapters but it is amusing in passing to record Professor Roget's famous 'cow test' which was often propounded in early English skiing literature. 'When in doubt,' he wrote, 'the ski-runner should ask himself: Are cows as I know them likely to feel comfortable when standing on this slope in summer? If an affirmative answer can be given in a *bona fide* manner the slope is not dangerous.' The Genevese professor must surely have been indulging in a gentle leg-pull, but many people failed to realize it.

If you are armed with the knowledge that you are heading unavoidably into avalanche danger your chances of staying alive are considerably improved, and they can be enhanced still further by taking a few simple measures. The most common precaution, and the one believed to be almost foolproof until recently, was the wearing of an avalanche cord. This is nothing more than a red cord about 25 yards long and about $\frac{1}{4}$ inch in diameter. One end is tied around the waist and the rest left to trail behind like a long tail. The principle is that part of the light cord will remain on the surface if its wearer is buried by an avalanche. The rescuers find the cord, pull it out of the snow until it goes down vertically, and then dig the victim free.

This invention—if it can be given such a grandiose name—dates from the early 1900s when a Bavarian mountaineer called Eugen Oertel thought of it. Only since the early 1970s has its efficacity been in doubt; a private foundation, of which we shall hear more in a later chapter, recently financed some tests with avalanche cords attached to small sandbags and found that quite seldom was any part of the cord left on the surface of the avalanche deposit. (One of the reasons is doubtless the settling of the snow dust that accompanies most dry-snow avalanches.) Nor could the foundation track down documented evidence of many people being saved by avalanche cords. My own interpretation of this latter point is that the ski alpinist who is canny enough to carry a cord, and use it when he senses danger, usually has enough respect for, and knowledge of, avalanches to avoid getting caught in one.

As we shall see in a later chapter on rescue methods, there are now some sophisticated devices that a skier can carry to ensure that he will be quickly located in an avalanche. If these are not available, I would still use an avalanche cord for the following reasons: it is better than nothing; in poor visibility it helps members of a party of ski alpinists to keep contact, and at the same time distance, when climbing or walking; I have personally found that unravelling what for me are the inevitable twists and tangles of the rolled-up cord, and tying it around my waist, give me time for silent reflection, a valuable chance to appraise calmly all the factors governing a decision as to whether to venture on to the slope ahead or not!

When about to move on to a slope which you think dangerous, you should try to release an avalanche first by throwing rocks or lumps of ice. Jumping up and down near the edge or sending a member of the party on to it, well belayed on a rope from a safe place, may also be effective. If not, the party should move on to the slope one at a time while the others wait and watch until each reaches safety again. This is not always possible, and if more than one person must be exposed to danger at a time they should be well spaced out. A 50-yard gap between members of a party should be considered the minimum. Not only does this ensure that fewer people will be carried away by a possible avalanche, but spreading of the load on the slope in this way also reduces the chance of starting one.

In any event, there should always be one person in a safe place watching those in danger. His eyes must never leave the slope and if an avalanche occurs it is his duty to note carefully where the victim(s) were taken from and where he last saw him (or them). This information will be vital to the rescue operation. For the same reason, each member of the party should watch very closely the person ahead of him.

The strictest discipline is necessary in keeping open order on a slope as was shown by an avalanche accident in the Samnaun area of eastern Switzerland on March 17th, 1958. A large party of German skiers, 21 in all, were touring with two ski-instructors. In the early afternoon the party was climbing across the southerly slope leading to the col above the Planer Salas. A ski-instructor was cutting the track and he had told the party to keep a distance of 30 metres between members. Admittedly, this was a somewhat meagre spacing, but sufficient in the circumstances because all was going well and the party was strung across the whole slope.

The leading skiers had already traversed the steepest portion of the slope, and were on safer ground beyond, when the rear marker, Dr. Gaertner, suddenly decided that the pace was too slow for him and forged ahead to overtake some of the skiers in front. He had passed five and was cutting a new track above them as he drew level with the sixth on slightly steeper ground. Suddenly, the combined weight of the two men caused the snow cover to fracture just above Dr. Gaertner's skis.

A small avalanche formed locally which carried away Dr. Gaertner and the three skiers nearest him. Then the fracture propagated upwards and across and the resultant avalanche affected the whole steep portion of the slope. Some of the 16 skiers involved managed to get away in the time-lag before the secondary and larger avalanche had formed. Others were only partly buried and were soon freed but Dr. Gaertner's action resulted in his own death and that of four others in the group.

Ideally the track cut in a dangerous slope should be as divergent as possible from any likely avalanche fracture line. In effect this means that the safest path to take on a slope is vertically up and down it in the fall-line. Goethe knew this fact and mentioned it in his book *Travels in Switzerland and Italy* of 1799. When describing the avalanche dangers of the Furka Pass and the route used by guides carrying chamois skins in winter he wrote: 'They go straight up the final slope. It is a safer route but more arduous.'

Following the fall line, though a useful tactic, is indeed more arduous; usually it means taking off your skis but it should be remembered that this alone is a safety measure because boots are less avalanche-provoking than skis.

If you are to keep your skis on in a danger area it may be worth while slackening the settings of the safety bindings so that the skis will come off easily if you are caught. On the other hand, you may feel that there would be a fair chance of skiing away from an avalanche on a given slope, and in this case slackened safety bindings might release at an inappropriate moment. This can only be decided on the spot, but in any case you *must* free your wrists from the loops on your ski-sticks if you are to have much chance of staying near the surface of an avalanche.

As you move into the danger area tread lightly, be constantly on the alert for the first sign of an avalanche and keep planning what you will do in those first seconds should one occur. And do not think, even if you are the last of the party, that you are necessarily safe.

In 1908, F. A. M. Noelty wrote an article telling skiers about avalanches and their avoidance, and ended with the following advice to anyone actually caught in one: 'Whatever happens keep your nerve and trust in Providence; the rest is beyond the disposition of man.'

Without wishing any disrespect to either Providence or Noelty, it is now possible to give more definite advice for improving your survival chances in an avalanche. Not that Noelty's advice is to be despised—indeed as far as it goes it is excellent—but it needs the addition of some practical hints because the situation is not entirely 'beyond the disposition of man'.

When the avalanche breaks there will be a split second, or occasionally several seconds, before the snow gathers momentum. Unless you are an exceptionally good skier, or the avalanche is one of wet snow, there is scant chance of being able to ski away from it. All in all, it is usually better to use that momentary pause after the fracture to look quickly around, determine the extent of the break and see whether there is anything like a rock or shrub below you to which you might be able to cling. If the break has occurred under your feet, or only just above them, a quick leap up-slope may land you in snow which is not avalanching. In any case, you should try to delay your departure by any means available—even driving a ski-stick into the snow. It is unlikely that you will be able to stop yourself being carried down, but every moment's procrastination gives you a better chance of remaining on top of the avalanche.

Once under way you must keep your mouth firmly shut and make strong swimming movements with arms and legs in an effort to stay on the surface. If you noticed, before moving off, that you were quite near the edge of the avalanche you should try to work your way in that direction.

You should bring an arm in front of your nose and mouth as the avalanche slows down because this can create a few valuable inches of breathing space if you are entirely buried. Once you have come to a stop you should make just one determined effort to free yourself by forcing upwards. Most people conserve their sense of direction in an avalanche but, if in doubt, the classic method of deciding which way is upwards is to spit and see which way the saliva runs.

If the initial attempt to break out is fruitless you should lie still and try to calm yourself, admittedly easier said than done but it is important because fear increases the oxygen consumption of the body. In some recent tests it was shown that fear can put up the breathing rate from the average 15 breaths per minute to 35 per minute, and occasionally to as many as 68 per minute. Nature's purpose is of course to provide the extra oxygen required by the body to fight, or run away, but as you cannot do either of these when buried in snow you might as well try to keep calm. You should only waste breath by shouting for help if rescuers sound very close, but even then the chances of being heard are slight.

All the above has assumed that you were in the break-away zone of the avalanche and that the snow, even if it started as a slab, broke up and flowed as loose snow. If, on the other hand, the slab was very compact and remains in large blocks it is sometimes possible, as the avalanche starts, to cling to a block and ride down on it. In this connection, the first edition of this book mentioned an American who claimed to have ridden a block of slab 3,000 feet down on to a frozen lake and enjoyed the ride. He was unidentified, and the report on the event was laconic to say the least. However, the person concerned read the book and was prompted to write to me; he was William

Putnam, a person who did some important snow and avalanche study in the U.S. around 1950, but who preferred to remain anonymous when describing his 3,000-foot ride. It seems I can do no better now than to quote from his letter to me, for his own words portray very clearly the development of the accident and its fortunate outcome:

'In the winter of 1943, a group of us were taking a busman's holiday from a mountain training centre and set out to climb Homestake Peak, one of the 14,000 footers in the Colorado Rockies. At an altitude of approximately 12,000 feet, we came gradually on to an increasingly obvious wind-slab area. Determining that it was probably safe for us on foot if we ascended the fall line, we continued. However, several hundred feet above us on this rather wide slope, a group of skiers who had been climbing up a neighbouring ridge commenced to traverse the slope. At this point, we unroped because we didn't want to be snarled up in what seemed to be developing into a messy situation, and we began to trail our avalanche cords and pray. The skiers did cut the slope off, but we were able to dig in a goodly bit so that the bulk of the avalanche passed us before we were finally pulled loose. At this point, we hopped on to the tail blocks, which were fortunately rather substantial, and rode out on to Homestake Lake where the bulk of the avalanche ended up. The skiers were badly mauled, but we extracted them, cursed them a bit, and then went back to camp.'

If an avalanche comes down on you from above there is a far more likelihood of a deep burial. It is important to have an arm in front of your face before the snow hits you, especially if there is much airborne-powder snow which could be forced into the lungs. You will be so bowled over that swimming will be difficult, but an attempt should be made none the less.

From time to time skiers and climbers have had fantastically lucky escapes from avalanches. Perhaps the outstanding one occurred in 1952 on Castor, a peak near Zermatt, when James Riddell and a party were climbing up from the Bétempts Hut at 8.30 one beautiful morning. They heard a crack and looking up saw an enormous mass of snow and ice blocks plunging straight at them. The wave was at least 100 feet high.

There was not a thing to be done in those seconds of waiting so they just cowered down with their backs towards the approaching avalanche. The sky went dark, the roar filled their ears and at each instant they expected to be crushed like flies. Snow dust swirled around them, but still nothing struck them in the back. The roar diminished, silence followed and they stood shakily to their feet. To their astonishment they saw that, just above them, an enormous crevasse had appeared where previously there had been none. It was several seconds before they realized that a snow-bridge over the crevasse had collapsed and that the glacier had swallowed the whole enormous avalanche. James Riddel reckoned that had the party been six minutes further on its way it would have been beyond the protection afforded by the hidden crevasse. Riddell recalled that some trouble he had had with a ski-binding had delayed the party by about that length of time.

No chapter on avalanche death and survival could exclude the sad story of Edy Bridy. At 09.20 on April 18th, 1962, an enormous spring avalanche of wet snow thundered down from the slopes above the Great St. Bernard road about a mile from Bourg St. Pierre. The new covered road to the tunnel was in construction at the time, and the avalanche swept across it and down into the valley below where it destroyed a large canteen building belonging to a company engaged on a hydro-electric scheme. It then piled snow 70 feet deep in the bed of the River Dranse.

Someone realized that, although the canteen was empty at the time, Edy Bridy, a 25-year-old border guard, was on duty in the area. And sure enough it shortly came to light that he was missing. The prospects for a successful rescue in such hardened, debris-filled snow were not good, but a team of 40 men was soon at work Later, an excavator was brought in to help.

Bridy was buried some 20 feet deep, trapped between some of the timbers of the canteen. He could hear the rescuers at work and was reassured by their presence—until he heard the excavator. As he listened to it crunching its way through the snow he was convined that the bucket would cut him in two. Overcome with terror, he managed to work his pistol from its holster and let off several shots into the snow, but they went unheard.

As night fell the rescuers worked feverishly on by the light of flares, and at 01.00 hours they made the first contact with Bridy. He had remained conscious throughout in his cramped position among the timbers. At 01.32, after more than 16 hours burial, he was freed and after a change of clothing he was taken to Martigny hospital, where he arrived at 04.30.

His injuries were slight, and later in the day his family visited him. He seemed normal and chatted with them without showing any signs of shock. On April 20th he made out his report of the accident and by April 21st, when visitors again came to see him, he had apparently made a complete recovery.

On the 22nd his superior visited him and Bridy talked to him happily, asking to see photographs of the avalanche. But at 20.30 hours he suddenly seemed nervous and upset and asked not to be left alone. He was given a sedative and he went to sleep. At 21.25, a nurse was passing the door of his room when she heard a rattling in Bridy's throat. She summoned a doctor at once but Edy Bridy was already dead.

The autopsy showed 'nothing special' and one can but assume that he was killed by delayed shock some four days after the accident. The legend is already growing in the area that Bridy's body had taken up the same position in bed as during his 16-hour burial. As the local people now say: 'Poor lad! He would have suffered less had he died immediately in the avalanche.'

8
Rescue Organizations

In the 10th century A.D. a certain wealthy young man decided on the eve of his marriage to renounce the pleasures of the world and to devote his life to helping those in need. History does not tell us anything about his fiancée and it is agreeable to imagine that he renounced a veritable Venus for the sake of his beliefs—but she could equally well have been a virago, and frumpish to boot. If she was, and the fact influenced his decision, the mountain world has cause to be grateful to her, because the man concerned was Bernard of Menthon, later to become St. Bernard. He joined the Augustine Order and by A.D. 962 he had founded a monastery at an altitude of over 8,000 feet on what was then called the Mount of Jove. The purpose of the monastery was to help travellers and pilgrims in difficulty in the mountains and he named it after St. Nicolas. Subsequent to his death it became the Great St. Bernard Hospice.

It is not surprising that Bernard has become the patron saint of Alpinists, for he founded what amounted to the first mountain rescue organization, in the sense of a group of people willing to go to the assistance of strangers in the mountains. It is hardly necessary to point out that the scope of this chapter does not include those rescue groups formed spontaneously by the men of a village after an avalanche or other disaster has struck.

After the Great St. Bernard, further hospices were built by the Augustine Order on the Simplon and Lesser St. Bernard Passes. Other orders followed the example until most of the major passes of the Alps had a hospice at the summit. Today it is easy to forget the tremendous and worthwhile work which these hospices carried on until the boring of the main Alpine tunnels around the end of the 19th century. For example, in the 18th century, some 15,000 people a year were crossing the St. Gotthard Pass, and at the Great St. Bernard Hospice it was not at all rare to serve 400 meals a day.

During the winter a number of travellers died on each of the main passes. Father Lorenzo, who spent 17 years at the St. Gotthard Hospice, reported in 1783 an annual average of three to four deaths from avalanches and freezing. And in the early 18th century, Prieur Ballalu of the Great St. Bernard wrote: 'Hardly a year passes in which one or more people do not die here; some are taken ill; others are surprised by avalanches in the mountains, or are overcome by cold and freeze before arriving here, or being carried here.'

These deaths occurred despite the best efforts of the monks to guide and rescue travellers; without those efforts the death roll would doubtless have run to hundreds each year. Every winter morning for centuries, a monk or hospice servant would set out in each direction from the St. Bernard Hospice, one towards Bourg St. Pierre and the other towards St. Rhemy. They guided travellers down, and most people who wished to come up would wait in a shelter near the bottom of the Pass for the guide to arrive and escort them to the summit. It is not known when this practice began but it was certainly long before the drawing up of the 1436 Constitution of the Hospice, in which the duties of the Order are defined. Among the edicts in the Constitution is one telling the monks that they are to go out each day between November 15th and the end of May to meet travellers from both sides of the Pass. But we know that this edict was not a new idea because it ends with the words: 'as they have always done.'

So it is that from well before 1436 until 1885, not a single winter's morning passed, regardless of how vicious the weather, on which men did not set out from the St. Bernard Hospice to guide travellers, or to rescue those in trouble. After 1885 their sorties were not always necessary because in that year telephone communications were installed to the Hospice, the first telephone in Switzerland incidentally. From then on, travellers were able to contact the Hospice before setting out. They were ordered not to attempt the journey in the worst storms and the Hospice personnel, in the knowledge that no one was in danger on the Pass, did not need to go out. In normal weather, however, they were still very busy because by then 25,000 per annum were crossing the Pass, a figure that only began to diminish after the opening of the Simplon Tunnel in 1906.

Who can know how many lives were saved by those dedicated men following St. Bernard's example throughout the centuries? With their famous dogs, the forerunners of today's avalanche dogs that are trained to locate humans buried in snow, they carried out thousands of searches through blizzards and darkness for those who had left the valley too late or strayed from the path. They rescued hundreds of people from avalanches, and frequently they found exhausted, frost-bitten travellers whom they carried to safety, often forcing themselves to the limit of their own endurance in the effort. We shall never know precise details of the rescues they carried out because no records were kept: the saving of life was for them the normal course of duty and it would have been unthinkable vanity to have made a note of their successes.

But duty cost many of those men their lives, mainly in avalanches. Between 1810 and 1845 alone, avalanches killed 12 men from the St. Bernard Hospice. Yet, without concern for their own safety, the Hospice personnel have brought help and rescue to anyone and everyone, from princes to highwaymen. Quite recently an avalanche killed a monk while he was guiding a group of Italian smugglers off the normal path. If one expresses surprise that such people are also helped, the Prieur says with a smile: 'Our duty is to all travellers. Why they are travelling is no concern of ours; and in any case the taxes and tariffs which make the smuggling of tobacco into Italy so profitable were imposed by man, not God!'

Not only did St. Bernard found the first mountain rescue organization, but he also founded it in traditions of endeavour and self-sacrifice in the aid of all and sundry. And these traditions still inspire the better rescue organizations today.

The tourist in the mountains nowadays has the comforting knowledge that, if he falls into difficulty, a group of trained and dedicated men with modern equipment can be summoned to his assistance. Early climbers had, in the main, to be self-sufficient in their rescue requirements as is well illustrated by the Birbeck accident of 1861.

John Birbeck was climbing with some of the most famous mountaineers of the epoch. He was but 19 years old and he had been entrusted to the Reverend Charles Hudson for his first season. The party had unroped to have breakfast on the Col de Miage, in the Mont Blanc massif, when Birbeck wandered a few yards away, slipped, and fell 1,800 feet down an ice slope. He would almost certainly have died had it not been for the Victorian bent for invention; the party had carried with it a collapsible rescue sled especially designed by Hudson. The runners formed his *Alpenstock* and the boards were carried, two apiece, by the other members of the party. Hudson collected them and soon had the contraption screwed together. It is remarkable that the party, with a difficult climb in prospect, had so encumbered itself, but the fact clearly illustrates that mountaineers of the time were well aware of the need for self-sufficiency in the event of an accident.

Birbeck was transported part-way down the mountain. Sir Leslie Stephen went to Chamonix to find an English doctor; Fox Tuckett went to the main valley to telegraph Geneva for a surgeon, and a guide went to enlist the help of some local men with a stretcher. After a tremendous struggle lasting all day they reached the valley with Birbeck and, although he was by then in a very weak state from shock and his appalling abrasions, he survived. It was an outstanding feat of rescue when one considers the few means at their disposal.

The first non-ecclesiastical rescue organization in the Alps was probably the one set up by the Duchy of Savoy in the Middle Ages. In winter, certain men were excused military service to form a group of guides and rescuers to assist travellers. They were called the 'Soldiers of the Snow'. Nowadays, however, with the phenomenal growth in tourism, the rescue network in the Alps is very comprehensive and many different units and organizations contribute to its efficiency. None of them specializes in avalanche rescue, but this sphere nevertheless plays a large part in the activities of each and every one of them. It is therefore worth examining the rescue network to see how it functions and to give an idea of the activities of each unit.

The widest-flung organizations in the Alps are those of the Alpine Clubs, mostly formed around the middle of the 19th century. They have small voluntary teams of rescuers in almost every valley. The head of the team is, without exception, an experienced Alpinist, sometimes a guide or sometimes the local doctor or schoolmaster. In any case he is a man with a feeling for the mountains and a willingness to help fellow-Alpinists in trouble. Rescue equipment is on hand locally and more comprehensive supply depots are located at central points. In time of need the small teams of the Alpine Clubs are reinforced by guides, ski-instructors and any other able-bodied men who can climb or ski.

The network of the Alpine Clubs is well able to cope in summer, but when the hundreds of thousands of skiers visit the Alps in winter, with all their arrival entails in the way of broken limbs and avalanche accidents, additional organizations are necessary. The government authorities will not license a funicular or cable-car unless it has adequate rescue personnel and equipment at its disposal. Some cable-cars satisfy this requirement by taking on a few extra employees to form their own small, independent rescue team, while in other places a single and larger organization, supported by every lift, funicular and cable-car in the area, is responsible for all the rescue work.

Such an organization is the Parsenndienst at Davos and there can be no doubt that where the grouping of the lifts lends itself to a single organization, as it does in the Parsenn area, greater efficiency results. This is brought about by standardization of training and rescue procedure, as well as by the fact that money is available for the purchase of expensive equipment which can be stored centrally for use in the whole area.

The Parsenndienst features regularly throughout this book, but no apology is made for the fact. It is not, on the other hand, being insinuated that no other good rescue organizations exist—who can deny the excellence of units like the Bayrische Bergwacht (Bavarian Mountain Guard)? But the Parsenndienst has won a world-wide reputation for its efficiency and its methods. It does not operate a service for climbers in summer, but in the field of skier rescue in winter it stands pre-eminent. And the enormous increase in the popularity of skiing since the war has made this the most active form of mountain rescue, even though in the majority of calls it is a matter of broken bones rather than the usual life and death affair of a climbing accident.

Nevertheless, with avalanches now claiming over 80% of their victims among skiers, avalanche rescue and safety play a very important part in the activities of a skier rescue service. The International Commission for Alpine Rescue has established guidelines for the functioning of these services. It is a mark of the recognized qualities of the Parsenndienst that, with regard to avalanche rescue, the guidelines are based to a very large extent on practices first developed and used by the men of that service. It is interesting, therefore, to examine the Parsenndienst closely and see how it functions in its various activities, and particularly to see how it answers to the most vital of all its tasks—an avalanche rescue.

The Parsenn is a very large skiing area with about a dozen lifts, cable-cars and railways feeding no less than 93 miles of marked ski-runs, of which 45 miles are used intensively. Some of the runs reach the valley at points in excess of 15 miles from Davos and skiers then return by train. To watch effectively over an area this size the Parsenndienst employs a manager and about 30 men, and has running costs of well over $100,000 per winter, of which about 15% is recovered by fees charged for rescues.

Sixteen of the men are patrolmen trained to give first aid to injured skiers and transport them to hospital on sledges, while the remainder of the staff spend most of their time working on the ski-runs.

The Parsenndienst was one of the first organizations to try to reduce skier accidents by levelling bumps and filling in holes on ski-runs, first with shovel gangs and now, of course, with snow cats. Despite all their best efforts, and despite safety bindings, there are still between 500 and 600 injured skiers to be transported to the valley each season. However, only about a third of the organization's total effort is devoted to actual rescue; the larger proportion is taken up with safety and the prevention of accidents. In this field comes the aforementioned bringing down or controlling of avalanches with explosives after a snow-fall, and for this activity every available man is needed.

From small beginnings in 1927, the Parsenndienst has grown into a sizeable organization. Its members are justly proud of the fact that they have transported more than 10,000 injured skiers to the valley without a single serious incident. Most of the credit for the performance of the organization must go to Christian Jost, its founder and for many years its head.

Jost developed his interest in mountain rescue during the First World War when he was with the Alpine Troops. In 1914 he and five other soldiers were sent out with the guide Hans Kaspar to search for a well-known mountaineer, by the name of Schaufelberger, who was missing in the Bernina range. The party ran into bad weather and they were all frost bitten, with the exception of Kaspar who had stuffed newspaper between his two pairs of socks.

When they finally reached a hut, Kaspar mixed salt and snow in a bowl and told the soldiers to stamp around in it. Christian Jost related that the pain was excruciating but he clenched his teeth and persevered. He was the only man among the six who did so however—and he was the only man not to suffer any amputations later.

It would not have been surprising had such a painful introduction to rescue work put Jost off for the rest of his life, but, on the contrary, something about the experience appealed to him. After the war, when he returned to his job as schoolmaster at Davos, he promptly joined the Swiss Alpine Club rescue team there and later became its leader. By 1927 he was also President of the Davos Ski Club and he founded and ran a rescue service for the Club. This service was, in effect, the beginning of the Parsenndienst.

With the completion in 1932 of the Parsennbahn, the cable railway from Davos to the Weissfluhjoch, more and more skiers flooded into the area. Not everyone was as quick to realize the importance of an efficient rescue service as was Christian Jost and this was especially true on the occasions when he was requesting additional finance for the service. Then, as now, the cost of the Parsenndienst was paid by the various interested bodies in the area—that is to say the railways, ski-lifts, public interests associations, communes and so on. But in the early days there was much wrangling as to the total amount of money to be spent and the proportion to be paid by each. Jost, although all the while carrying on his job as schoolmaster, campaigned ceaselessly to improve the Parsenndienst and to make people aware of its value. The measure of his success is that, in recent years, whenever extra finance has been required it has been provided with hardly a murmur.

New headquarters for the Parsenndienst were built at the Weissfluhjoch in 1957 and like the Federal Snow and Avalanche Research Institute they form part of the Parsennbahn station complex. These headquarters are very fine, light and pleasant with the usual pine panelling. There is a large staff-room, simply furnished with tables and chairs, where the men wait until they are called to an injured skier. In an alcove off the staff-room is a kitchen where the men prepare their own meals.

The office contains the main receiver/transmitter of the radio apparatus which the Parsenndienst has used since 1959. Until then there was no 'walkie-talkie' system which would function well in mountainous terrain without relay stations on prominent peaks, but the sets now in use are very satisfactory. They run on batteries charged from mains electricity, are light and robust, and if a patrolman finds himself in a 'dead spot', a movement of very few yards in any direction usually re-establishes communications. These radio sets have proved their value time and again at avalanche rescues, during search actions, avalanche blasting and bad-weather patrols.

The ski-runs of the Parsenn are well equipped with telephones for the reporting of accidents. These telephones are connected straight to the exchange in Chur, the cantonal capital, where the operator will quickly find an interpreter for most common languages and pass a message on to the Parsenndienst if the caller does not speak German. There are three telephone lines to the headquarters and it has been found that in times of stress, say when an avalanche rescue is in progress, this is the minimum requirement.

The headquarters also include stores for equipment such as sledges, splints, avalanche sounding-rods, avalanche shovels, medical supplies, artificial respiration pumps, petrol-engined lighting plants, explosives, a 3-inch mortar, a bazooka and the many other items needed by a good skier safety and rescue unit.

There is a well-furnished sick room, and a bunk-room because the headquarters is manned day and night throughout the winter. It is surprising the number of night sorties that take place to hunt for people reported missing. They are usually found to have strayed from a run, or to have started out too late in the day and been benighted, but an amusing incident took place before the war in which neither of these things had happened.

At about 21.00 hours a hotel reported that an Englishman, whom we shall call Mr. Black, had not returned. The usual procedure when this occurs, and there is no clue as to where the person might be, is to begin the search by ringing up literally hundreds of hotels, bars and cafés. This was duly done, but with no success in locating Mr. Black. The staff of the Parsenndienst in the Weissfluhjoch headquarters were despatched to search some of the ski-runs and a special train was laid on to take up reinforcements.

By dawn, with the area searched as well as an area of that sort can be, the quest was called off, and as the first trains began carrying skiers to the Weissfluhjoch a notice was displayed asking Mr. Black to report to the Parsenndienst. Sure enough, after a while, a man arrived, said he was Mr. Black and asked why he was wanted. It was explained that a search had been in progress for him most of the night and that the Parsenndienst would be most interested to know where he had been.

Mr. Black looked acutely embarrassed and eventually explained that he had heard his name called out in the bar where he had been early in the evening, but that he had wished to remain incognito because of the company he was keeping. He had later gone on to a strange bed; but his pleasures there proved expensive because he was presented with a bill for the search. The moral of the anecdote is not for a skier to deny himself pleasures, but always to advise his hotel by telephone of a safe descent to the valley if he does not intend to appear in person.

The development of the Parsenndienst doubtless required the leadership of a man of both vision and character. Christian Jost had both in abundant measure and coupled them with great organizing ability. He directed the service until he was in his early 70s—when I first knew him—and no longer able to ski, but his experience was so wide, including no less than seven avalanche burials, and his memory so good, that his physical inabilities were of small account.

From his office in Davos he directed affairs with great efficiency by spending many hours each day on the telephone. He knew exactly what was going on at all times and his men brought him detailed reports on the condition of the runs. With his memory for every rock and bump in the Parsenn area, he could build up a precise picture of the situation.

He was a bear-like man in his last years, heavy-jowled and ponderous of limb, but his mind was as lucid as ever. It was a delight to hear him explain something, or lecture on rescue techniques. He had the valuable gift of inspiring both respect and affection in his men, many of whom he had known since he taught them in school. From their remarks it emerged that he was one of those rare school-masters with sufficient character to maintain discipline with but rare recourse to threat or punishment, a patriarchal figure who taught with great patience and understanding.

Jost smoked at least 40 cigarettes a day which, in part, accounted for the rasping voice with which he growled his orders over the telephone each morning. These orders were so detailed and clear that each man knew precisely what was expected of him; hence there was seldom trouble over work incorrectly carried out. Indeed, some people claimed that Jost was so explicit that he killed initiative among his staff, but this was not a fair criticism. The patrolmen took many decisions; they carried much responsibility, and Jost was the first to encourage a man to use his wits and own powers of judgement.

Christian Jost had one knack at which I repeatedly marvelled. He would sit in his office contemplating the snow conditions and the weather, and then he would suddenly declare that a certain run was to be closed at once because of avalanche danger. Time and again an avalanche came down a few hours later where he had predicted; or the patrol which he sent out released one. He thought, of course, in terms of temperature changes, the snow structure, the direction of recent winds, and, from his records of avalanches in the area, he knew what slopes were still laden with snow. It was a remarkable achievement, however, to be able to predict in this way, though Jost dismissed it with a chuckle and said that it was just experience.

Jost brought great humanity—behind his somewhat autocratic exterior—to his work with the Parsenndienst, not only in his concern for the well-being of injured skiers brought off the mountain, but also in his concern for his staff. When his men were out on dangerous patrols, he had a telephone receiver placed near the radio in headquarters so that he could follow every step of the action as they reported to base. He could not rest easy until his men were all safely back.

Jost retired from the Parsenndienst in 1965 and died in 1967. His death was a loss to the mountain rescue fraternity and brought great sadness to many, including me, for I had come to admire and love him during the winters I worked under his orders. And I have fine memories of the many convivial evenings spent with him when I imbibed of his vast experience—and also of his excellent wine. Luckily his place in the Parsenndienst was eventually taken by another extraordinary person, Nic Kindschi. Even if much less patriarchal than Jost, he still has great qualities of leadership, expertise and humanity, and so the Parsenndienst continues in its traditions.

Some skiers may have been alarmed by the aforementioned fact that 500–600 accidents occur each winter in the Parsenn area alone. On the other hand, they may be reassured to know that this only represents one injury for approximately 1,000–1,200 ski descents carried out. They may also be reassured by knowing, in brief, what happens if they are unfortunate enough to have that chance in a thousand come up while they are in the Parsenn area.

As soon as it becomes apparent, from the one-sided telephone conversation in the Parsenndienst staff-room, that an accident has occurred, the patrolman whose turn it is to go grabs his rucksack off the shelf, and passes it to a colleague. He picks up coat, hat, gloves and skis and starts towards the door, only delaying long enough for the duty telephonist to tell him where the accident is and what sort of injury he is going to. In the meantime the colleague has filled the hot-water bottle and tea-flask in his rucksack. While the patrolman is putting on his skis the colleague ties the rucksack to the sledge and in an instant the patrolman is hurrying down the mountain in a series of rapid 'schusses'.

Dependent on the length of journey, he will be beside the injured skier in a time varying from 5–20 minutes after the call was received. On the occasions when there is a longer delay the usual reason is incorrect reporting of the accident location. The patrolman then rings back to say that he has had an empty run, a check is made and a second patrolman despatched. Admittedly, there can be crisis days from time to time. There was one in February 1964 when there were 11 injuries (not in itself a high total because the record is 23 in a day), but 7 of the 11 came within a 42–minute period. The resultant strain on the organization caused one patient a wait of 45 minutes, but he was in a warm hut and warned of the likely delay.

When the patrolman reaches the patient he makes a quick diagnosis. The commonest injury is a broken tibia and/or fibula, though far more dangerous and unpleasant injuries sometimes occur. The patient is given a pain-reducing injection or tablets, according to need, and then drinks a mug of hot tea while the injury is attended to. This is usually a question of splinting a leg and applying a little traction to prevent the bone-ends jarring on the journey to the valley. The patient is then loaded on to the sledge with foam rubber and inflatable cushions under him, and a hot-water bottle is placed on his stomach. Enveloped in a wool-lined canvas wrap, he is firmly lashed to the sledge with one of his skis on each side of him.

During these early stages of the rescue, the patrolman talks quietly and sympathetically to the patient. He goes about his work with such reassuring calmness and dexterity that there are hundreds of patients' letters in the Parsenndienst archives, many of them from doctors, praising the way they were handled.

On the journey to the valley, the patrolman stops frequently to enquire after the patient's comfort and at the first opportunity he rings up headquarters to tell them to which doctor or hospital in the area he wishes to be taken. Headquarters then order an ambulance to meet him at the end of the ski-run. The patrolman always accompanies the patient until he is on an X-ray table before packing up his equipment and returning to headquarters. The charge for this service averages $20–25 but will be more or less according to the distance the patient is transported.

From this account it will be obvious that not only must a patrolman be a fast and reliable skier, but he must also be trained in first-aid and the use of a rescue sledge. It will have been noticed that the patrolman carries through alone the task of transporting an injured skier, although in many skier-rescue units it is common practice for two men to transport a patient. The matter depends entirely on the type of rescue sledge, and there are two different sledges in common use throughout the Alps. The first type is the Akja, originally from Scandinavia, and the other is the so-called Kanadier (Canadian) sledge.

The Akja is a boat-like sledge with a set of shafts at each end. Modern models are usually made of aluminium and they have found a good market, particularly in Austria. Nevertheless, the Akja has the serious drawback of requiring two men in its use—firstly to lift the patient into the sledge and then to bring it down. It is also somewhat slow because the usual method of coming down a slope with an Akja is to make a series of traverses with the two patrolmen turning round in the shafts at the end of each run. On the other hand, the two sets of shafts give the Akja the advantage of being easily carried over very rough terrain, or over snow-free patches in spring.

The Canadian sledge is made of wood, flat, but with a curled-up front like the sporting toboggan of North America. This was originally the type of sledge used by trappers in the Arctic for carrying their furs. How it was brought to the Alps and became the commonest sort of sledge for mountain rescue is intriguing and, in certain details, a mystery.

Sometime towards the end of the last century a Canadian, or an

Englishman who had spent some time in Canada, brought a sledge of this basic type to Lenzerheide in Switzerland for his own sport and amusement. Exactly who this man was cannot be determined but he left the sledge behind and never reclaimed it. Sometime in the early years of this century, a skier at Lenzerheide broke his neck and the sledge, which had been gathering dust for years, was used to bring him to safety.

A Swiss engineer called Lindenmann, who later developed a number of devices for rescue and safety in the mountains, saw the potential of the sledge for rescue purposes and improved it by making it divisible into two sections for ease of carrying. (This was of course in the days before ski-lifts so a rescue involved carrying the sledge up to the patient.) In 1924 this modified sledge was seen by the head of the Swiss Alpine Club rescue team in Davos and he promptly asked Lindenmann to make three for him. The Davos Ski Club followed suit.

The Canadian sledge underwent development and improvement from then on, most of it at Davos under the auspices of Christian Jost. At the outbreak of war the Swiss Army requisitioned 12 of the Parsenndienst sledges and their experience with these later led to military purchases by the hundred. With their low centre of gravity, easy loading and requiring only one man to bring them down, they were found to be ideal.

Today's Canadian sledge is a familiar sight to most skiers. It is usually painted scarlet and this explains why Anglo-Saxon skiers usually call it a 'Blood Wagon'. It is about 6 feet 6 inches long with a single set of shafts at the front. A length of chain is attached at the base of the shafts and the chain loop so formed trails under the curved-up front of the sledge. This is the brake chain, and when the patrolman wishes to slow the sledge he pushes downwards on the shafts which forces the chain into the snow. So effective is this brake that even a small patrolman can bring a heavy patient down the fall-line. In very steep places the patrolman turns his skis across the slope and slides sideways to assist braking, but the majority of slopes can be taken straight. This makes for quick transportation of the patient to the valley even if the patrolman does not travel very fast with the sledge. Indeed, in a number of competitions between rescue organizations using Akja and Canadian sledges, the Canadian has always won convincingly.

To bring a patient down on a sledge is a knack and occasionally a beginner in training has difficulty. This happened once in the Parsenndienst when a senior patrolman had volunteered to be 'patient' during a training session. He thought the job would be restful, but the trainee fell and let go of the sledge which trundled away with the patrolman lashed helplessly to it. Someone overtook it and flicked a shaft so that it capsized and came to a halt. The patrolman was smothered in snow and indignant, to say the least.

The training of a patrolman is spread over many years. He begins by working with the shovel gangs improving the ski-runs and eventually, if his character and abilities recommend him, he will be promoted to assistant patrolman. Jost was once asked what training a patrolman had and he replied: 'Six or eight years of snow shovelling.' The principle underlying this cryptic answer is that any man of ability who will put up with the tedium and low rewards of shovelling snow, winter after winter in order to become a patrolman, has a real interest for patrolman's work.

In effect, his real training begins as an assistant patrolman. He usually remains at this level for four to five years and during this time he will attend first-aid courses, learn to handle a sledge and be initiated into the tricky job of controlling avalanches with explosives. In addition, before he qualifies as a patrolman he must be able to take charge of an avalanche rescue. The call could come at any moment and, in fact, the whole Parsenndienst is geared to meet the eventuality; even the most junior member of the staff is trained to react efficiently within his capacities. (We shall see in the next chapter how an avalanche rescue is put into motion.)

Once qualified a patrolman returns winter after winter, and this results in slow promotion within the Parsenndienst. The work is hard with much responsibility and frequent exposure to avalanche danger. A patrolman works a 10-hour day with only one free day a fortnight. His pay is about the same as that of a labourer, plus whatever grateful patients give him. In fact, the patrolman's earnings are far less than that of a ski-instructor for many more hours work a day. On the other hand, a patrolman's compelling interest in his work largely compensates for any deficiency in material reward.

Most of the patrolmen of the Parsenndienst are masons, painters, builders and farmers in summer. They are simple men, but men of courage, integrity and devotion to duty. A few years ago, one called Jakob Joos was hurrying to an injured skier when he failed to see a bump in the blinding snowstorm, fell and broke his own leg. He was more than 300 yards above the girl he had come to collect. He told a passing skier to telephone for two more patrolmen and then he made his way painfully to the injured girl. By the time help arrived he had her leg splinted and she was on the sledge ready to go. He had accomplished this crawling about on his knees with his broken leg dragging in the snow.

On another occasion, Paul Sprecher, a gay and humorous little man who was one of the first patrolmen, broke his leg while actually transporting someone on a sledge. He lay quite still for half an hour with the sledge partly on him rather than take the slightest risk of losing control of it if he tried to make himself more comfortable.

Of course, there have been many funny episodes in the history of the Parsenndienst, most of them involving Jost who had a fine sense of humour. One of the funniest occurred a few years ago when a patrolman went to collect a German skier who had broken his leg. When they reached the Wolfgang Kulm, a hotel in the valley at which several runs terminate, the patrolman telephoned Jost, told him where he was and said:

'My patient wants to split a bottle of champagne with me. Do you mind if I break the rule and drink on duty?'

'I thought he had a broken leg,' said Jost.

'Yes, he has.'

'Then take him to the hospital or a doctor at once,' said Jost.

'He says he doesn't want to go.'

'*Her Gott Sakrament!*' snarled Jost. 'He must go!'

'He won't,' said the patrolman, 'and he says he's travelling home tonight.'

There was a long silence at the other end of the line, and then Jost began to make the noise he invariably made when he was deep in thought—a sort of soft, windy whistle through pursed lips. Suddenly his voice exploded into the telephone again:

'*Ja*, then the man's got a wooden leg and he's broken it! Get on and drink the champagne!' And he hung up, laughing.

Since then, there have been a number of broken artificial legs in the Parsenn area, and the patrolmen are quite adept at carrying out running repairs in the workshop; but it was never possible to fool Jost over the phone again, even for a moment.

* * *

We have now examined the rescue organizations as they exist locally in the Alps, and each of these units can of course be helped by national institutions like the army, ski school, border guards, police, etc. in time of emergency. Additionally in Switzerland, however, there exists the *Schweizerische Rettungsflugwacht* (Swiss Aerial Rescue Guard). This organization is run on a national basis but will also help outside Switzerland if called. It is complementary to the local organizations and it will only make its men and aircraft available in reply to a call from one of them. It is supported to a large extent by charity and commercial companies because its operating costs cannot be covered by the payments for rescues carried out.

The *Rettungsflugwacht* in its present form was only founded in March 1960, but its origins go back to 1952. It was in that year that Hermann Geiger realized his dream by showing that a fixed-wing light aircraft could land and take off in safety on glaciers, snow-fields and tiny meadows high in the Alps. Until then, the only landings in the mountains had been isolated exploits for which large expanses of level snow were chosen, natural airfields in fact.

Hermann Geiger, a pilot and mountaineer from Sion in the Rhone valley, studied the problems of light aircraft operation in the Alps for a number of years before putting his ideas to the practical test. His first landings were on glaciers, where he found little difficulty, but he later developed a technique for landing in confined spaces. He would land upwards on short and quite steep slopes, pulling the aircraft to the top on the engine and then swinging it to one side so that it could not slip back. Taking off downhill added to his momentum and much reduced the necessary length of run.

This sort of Alpine flying demands the very finest of piloting skill coupled with a mountaineer's sense for wind and snow-surface conditions. Geiger quickly became something of a national hero, but he showed great self-discipline and maturity of judgement during the early years of his Alpine aviation. It would have been all too easy to overreach himself in attempting ever more audacious feats until the dire prognostications of the many sceptics were proved correct. Instead, with the canniness typical of mountain people, he progressed with caution until he had built up a record of hundreds of incident-free Alpine landings and had thus proven that his methods were safe and practicable.

His first commercial undertakings were the delivery of provisions and building materials to mountain huts but he followed these with rescue flights to collect skiers and climbers injured at high altitudes and a long way from ski-lifts and cable-cars. In May 1954, he evacuated five injured avalanche survivors and a dead body from Mont Calme, in the Grand Dixence area, in conditions of such severe avalanche danger that the rescuers had been wondering how they were to transport the survivors to the valley without further accident.

Geiger initiated other pilots into his techniques and Alpine aviation expanded fast. Helicopters with ever-higher service ceilings have been developed over recent years and these too are now pressed into service for Alpine rescue. The *Rettungsflugwacht* keeps numerous ski-equipped aircraft and helicopters tied by charter arrangement to ensure their availability with minimum delay when they are needed. The staff of the *Rettungsflugwacht* works voluntarily and part-time. It includes 65 pilots, 20 doctors, 11 parachutists, 30 ground support rescuers and 20 airfield ground staff. They operate from eight airfields spread throughout Switzerland and supplies of the most modern rescue equipment are stored at these points.

Immediate contact with the *Rettungsflugwacht* is obtained by dialling 11, the Swiss national emergency number, and asking to be put through to the headquarters in the control tower of Kloten Airport, Zürich. From there the *Rettungsflugwacht* establishes with the local organization exactly what is required and the nearest airfield to the accident is alerted. In the case of an avalanche accident in a remote area, the first need is usually for an avalanche dog and a doctor. In the interests of speed it is quite usual for a helicopter to collect them direct from their homes. Whenever possible the doctor is experienced in tracheotomy and the passing of a tube into a person's lungs, because both of these may be necessary during attempts to resuscitate an avalanche victim.

Provided that the first landing at the site of avalanche accident can be accomplished in daylight, it is usually possible to keep helicopter communications open after dark with the aid of flares. This

was the case at the avalanche which killed 11 children above Lenzerheide in 1961. At that particular accident the *Rettungsflugwacht* were
able to transport rescue equipment from Zürich before it could be
made available locally, though this is an indictment of the Lenzerheide organization rather than special praise for the *Rettungsflugwacht*.

The majority of the aircraft that the *Rettungsflugwacht* calls upon
in moments of need are more usually engaged in transporting materials to mountain huts and construction sites and in carrying tourists
and skiers to the glaciers. It is only this day-to-day Alpine aviation
activity that allows operating costs to be kept low enough—and for
there to be sufficient aircraft available—for the *Rettungsflugwacht* to
exist. Yet from time to time, especially in Switzerland, pressure
groups lobby to have Alpine aviation banned because of the noise
and disturbance they claim that it creates. These pressure groups
usually concede that the *Rettungsflugwacht* should be allowed to
operate, but they seem to ignore the fact that there cannot be an air
rescue service without more mundane aviation in the Alps.

One can sympathize with the purists who wish to prevent the
noise of engines disturbing the 'high places'. On the other hand a
single jet fighter of the Swiss Air Force or a jet airliner passing
overhead on its climb out of Geneva, Zürich or Milan, creates
more disturbance than several light aircraft with their petrol
engines of some 150 horsepower. It is also unlikely that Alpine
flying will ever become cheap enough to attract more than a fraction
of the skiing populace. In truth, a far greater threat to the peace of
the glaciers will be the cable-cars constructed by future generations
and the crowds they will carry up.

It seems to me, therefore, that it would be wrong to pass legislation of dubious value whose side effect would be the disappearance
of the *Rettungsflugwacht*, an organization which carries several hundred injured climbers and skiers a year to safety. Alpine aviation
must be here to stay and the pressure groups would do better to
accept its advantages, as well as its drawbacks.

*   *   *

Future developments in the organizational aspects of mountain
and avalanche rescue seem to lie in improving the existing network

rather than extending it. It is unfortunately true that in recent years some avalanche rescues have been carried out with such deplorable inefficiency that the chances of saving a life were negligible. Bitterness and recrimination have already followed in sufficient measure so there is no need to state where and when these bungled rescues took place. But it is interesting to note that the commonest mistakes and failings are ones of organization. Because no one was directly responsible, no one has telephoned for an avalanche dog, or no one has made sure that vital equipment was available. For one particular rescue not a single avalanche sounding-rod could be found; at another, the head of the rescue organization called off the search after less than two hours and when more than half of the avalanche had not been searched even superficially by the dog present.

It is such occurrences that have brought about the recent attempts to standardize avalanche-rescue procedures. The International Commission for Mountain and Avalanche Rescue hopes that by making known the well-tried procedures of the efficient organizations, the less efficient can be persuaded to adopt them. Fortunately, it is possible to standardize to a large extent in avalanche rescue because each operation is in general similar, and the same problems are being faced. This is in direct contrast to a climbing rescue in summer which can involve many different types of accident and may take place in terrain varying from a vertical rock face to a crevassed glacier.

Obviously, however, even in avalanche rescue, one cannot standardize the conduct of the rescue actually at the site because conditions are bound to vary. But one can lay down certain guidelines and minimum requirements for equipment. And, most important of all, one can lay down certain key roles in an avalanche rescue; the roles can be named and the exact responsibilities of each can be clearly defined.

Details of the type of arrangement and equipment considered desirable are given on page 150 and it is sufficient merely to name here the main roles in an avalanche rescue, with a broad outline of the functions to be performed by each.

The dominant position is that of *Rescue Manager*. In the Parsenndienst this post is filled by its Head or, in his absence, by a senior patrolman or one of two volunteer *Reserve Leaders*, both of whom are experienced mountain guides and men of standing in Davos.

The Rescue Manager directs the operation from an office. He is responsible for initiating the rescue, and for the ordering of avalanche dogs, reinforcements of manpower, extra equipment, special transport facilities and the *Rettungsflugwacht* if necessary. He also keeps local authorities informed, deals with the formalities should the victim have died and answers queries from press and radio.

The Leader of the First Team, usually a patrolman, is responsible for a dash to the site of the avalanche with as much equipment and as many men as can be mustered without prejudice to speed of action. The first team may well consist of only three or four men but whenever possible they will have an avalanche dog, a doctor and a radio with them.

On arrival at the avalanche the Leader of the First Team becomes *Accident Site Commander*. In this capacity he runs the actual rescue operation in the field. He wears a bright yellow overjacket with the words 'Accident Site Commander' written on it so that there can be no confusion as to who is in charge when reinforcements arrive. His responsibilities are heavy for it is he who has to assess the situation, decide from the evidence where to begin the search and what methods to use. He has to reach his decisions in the knowledge that every passing minute reduces the survival chances of the person buried. He is also responsible for marking with flags the edge of the avalanche, the skier's entry track and disappearance point, and the areas of the avalanche already searched.

The Accident Site Commander has to keep a minute by minute account of the proceedings with a sketch of the avalanche showing the important features. These detailed records are kept so that in a rescue lasting a long time the continuity of the action can be maintained, even if fresh men are called in to replace the original teams.

Unless there is a shortage of manpower the Accident Site Commander does not actually help with the search but merely directs it through a loud-hailer. The radio-operator stays beside him keeping a direct communication link with the Rescue Manager. All in all, the job of Accident Site Commander is totally undesirable. One, or perhaps more human lives depend on his decisions and in the wisdom which follows the event it is always possible to suggest ways in which his performance could have been bettered.

The remaining roles of responsibility in an avalanche rescue are those of the *Leaders of the Reinforcement Columns*. In the Parsenn

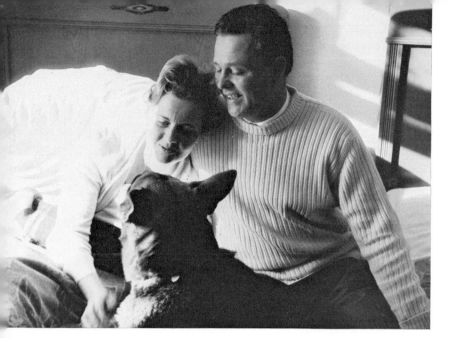

20. Dr. and Frau Kurz with Iso after he had saved their lives

21. A safety patrol of the Parsenndienst. The men are ready to go to the rescue of colleagues who are using explosives for avalanche control in the gully below. Their sounding-rods are visible and the patrolman on the left is in radio communication with the HQ and with the blasting patrol

22. The 3-inch mortar in use for avalanche control
23. The anti-tank bazooka under test for avalanche control

area these posts are filled by one of the Reserve Leaders, by a patrolman or by a ski-instructor. They lead the larger groups of rescuers to the avalanche, taking with them the equipment that the First Team could not carry because of the prime necessity for speed. Once the Reinforcement Columns reach the avalanche, they come under the control of the Accident Site Commander.

It is seriously to be hoped that standardization of avalanche rescue procedure can be achieved. Not only will it lead to greater efficiency within individual rescue units, but where several units work together at a major disaster, no confusion can arise if they all work on the same lines.

It will be realized from what has gone before in this chapter that an avalanche rescue can be very expensive. A team of 40–50 men may be at work for 24 hours at least and the *Rettungsflugwacht* may also be called in (as they are in some 50% of avalanche rescues in Switzerland today). Where those rescued can pay they are asked to do so, but as often as not they cannot. In such cases the local commune usually pays most of the bill—there is no question of pressing the survivor, or the family of the deceased, to pay. And the Alpine clubs also have a fund to cover the expense of a rescue in case the injured or their family cannot meet the obligation. Only in rare occasions can it be said that the people of the Alps fail to recognize their obligation to provide good rescue services for their guests.

9
Rescue Methods

Life and death, success and failure are seldom as close together as in an avalanche rescue. And one element dominates the whole operation—time, the most inexorable master of all. When a man is slowly suffocating, a few minutes, or even seconds, can dictate the outcome of the rescue, and we know already how few skiers are saved after a burial of more than two hours. Urgency is therefore of the greatest importance and there is a famous adage in mountain rescue to make the point:

'Without food a man can live for 30 days; without water a man can live for three days; but without oxygen a man will only live THREE MINUTES!'

The adage is not literally true because the human body can survive periods of 8–9 minutes without oxygen. However, permanent brain damage is so severe after three minutes that death is perhaps preferable. Nevertheless, the adage is useful as a means of impressing the need for speed in avalanche rescue and it is often used to introduce a lecture on the subject.

Few things are more frustrating and harassing than helping in an avalanche rescue. The area of tumbled snow may be quite small and you know that within those few hundred square yards a man is buried, possibly still fighting for his life. He may be but a few feet away, perhaps close enough to carry on a conversation with you in normal circumstances; and you know that as you try to locate him, each passing minute reduces the chance of finding him alive. It is an agonizing race against death, a race to be run fast but at the same time with clockwork planning and attention to detail. Unfortunately, it is a race which the rescue team more often loses than wins.

It is in many ways surprising that in this scientific age there is still nothing to equal a sounding-rod or a dog's acute sense of smell for locating a human body buried in snow. Other methods do exist, as we shall see, but they depend on the victim carrying special equipment and are not suitable for generalized use. Progress is being made in the development of devices which may meet all requirements, but it is unlikely that any of these will be functioning reliably for several years to come. These devices which are being developed will also be discussed later, but first let us deal with the means on which rescue organizations are at present forced to rely: sounding-rods and avalanche dogs.

Sounding-rods were mentioned by Strabo close on 2,000 years ago when he described the staffs carried by travellers in the Caucasus. He stated that when a traveller was buried by an avalanche he would push the staff up through the snow to indicate his whereabouts. It would seem from this that Strabo, quite literally, got the wrong end of the stick. It is rarely possible for an avalanche victim to push something up out of the snow, and it is far more likely that the staffs described by Strabo were carried so that those not buried could probe in the snow for those who were.

The first description of avalanche sounding-rods in the Alps was written by Nikolas Sererhard in 1742. Sererhard was a clergyman in the Prättigau valley, the scene, as it may be remembered from the first chapter, of many avalanche disasters. Today, avalanche sounding-rods exist in several different forms. The simplest type is merely a rod of mild steel about ⅜ inch in diameter and any length between 8 feet and 15 feet. There may be a loop bent into one end to act as a handle. The drawback of this type is that it is heavy and awkward to transport, but it is cheap and quite satisfactory for storage in mountain villages in case of disaster.

For use where the avalanche accident may take place some way from the equipment store, one of the more sophisticated, collapsible types of sounding rod is preferable. Colonel Bilgeri, a colleague of Zdarsky's, developed a rod of this type. It is of steel and in four sections that can be screwed together to make a total length of 11 feet 6 inches. The second main type is the one used by the Swiss Army. It, too, is in sections that can be screwed together but is of aluminium tube and has a normal total length, using four sections, of over 13 feet. Additional sections can be added for very deep sounding. (The rods slung across the backs of the Parsenndienst patrolmen in photograph 21 are of the Swiss Army type.)

There is a third type of collapsible sounding-rod, the Lindenmann rod, which has a slightly different application to those mentioned above. It is intended to be carried by skiers when touring, and with this in mind it has been designed so that lightness and compactness are its main features. It is hardly robust enough for a protracted and full-scale rescue operation, but it is invaluable for use by skiers who have survived an avalanche and are searching for their still-buried friends.

The weight of the Lindenmann rod is only 1½ pounds. Its total length of nearly 9 feet is made up of seven tubular sections through which a thin wire cable runs. Once the sections have been jointed they are held together by screwing a wing-nut at the top of the rod to tighten the wire running through them. Ideally, every skier on a tour should have a Lindenmann rod in his rucksack with his avalanche cord, unless the group is equipped with one of the modern devices to be described further on. At very least there should be one rod for every three members of the party, but then it must be remembered, when in avalanche danger, not to have the carriers of the rods too close together.

A search with sounding-rods has long since gone beyond the stage of indiscriminate probing here and there. It is now carefully organized and follows one of two patterns: fine (close interval) sounding or coarse (spaced) sounding. In both cases as many men as are available stand in line, shoulder to shoulder, and advance up the avalanche. Each man keeps his feet 20 inches apart and 10 inches from his neighbour's. In fine sounding, the rod is pushed down in front of the right toe-cap, then centrally between the feet, and then in front of the left toe-cap. This makes a long line of holes some 10 inches apart. The whole team then takes a 12-inch pace forward and repeats. This fine sounding gives near certainty of finding a body, but it is painfully slow. *To fine-sound an area of* 100 *metres by* 100 *metres (2½ acres) takes* 20 *men* 20 *hours when sounding to a depth of* 6 *feet.*

In coarse sounding, the men take up their positions in the same way but the rod is only driven down centrally between the feet, and the pace forward between probes is one of 2 feet 6 inches. *Coarse sounding gives only about* 70% *probability of striking a body, but* 20 *men can cover the* 100 *metres by* 100 *metres in four hours, sounding to the same 6-foot depth.*

The whole affair of sounding is fraught with conflicting interests: the longer taken over a search the better the chance of finding the victim—but the smaller the chance of his still being alive. And deciding on the scope of the search in the vertical plane also presents problems because it is much quicker to probe the top few feet of the avalanche than to probe the full depth.

Statistically, the best chances of a successful rescue are given by concentrating on speed rather than accuracy. Hence it is normal to begin by coarse sounding; and it has been quite recently postulated, on the basis again of statistics, that if coarse sounding has not revealed the victim at the first attempt, the best chance of a successful rescue will be given by coarse sounding a second time, rather than embarking on the very slow process of fine sounding. And at the outset the depth of sounding should be limited to 6 feet because a victim buried deeper than that will probable have died before sounding even began.

Of tremendous influence on the success of the rescue is the ability of the Accident Site Commander to make an inspired choice of priority areas of the avalanche in which to begin the search, rather than setting out to search the whole field. His decision can save many hours of sounding. In a small avalanche it is usual to search the fall-line below the point at which the victim was last seen, his so-called disappearance point. Sounding is begun at the bottom edge of the avalanche and proceeds up the fall-line.

If the avalanche is big, however, this search of the fall-line below the disappearance point could alone involve an enormous area, and the Accident Site Commander must then try to deduce where the victim has been deposited within that area. He will also have to try to make the same deductions if the victim's disappearance point was unobserved and the only available data is the skier's entry track at the edge of the avalanche.

Unfortunately, no pattern has been established for the way an avalanche deposits objects of different weight, shape and density—indeed it appears that no such pattern exists. Nevertheless, an American avalanche expert, Montgomery Atwater, did make some general observations on debris distribution when a large avalanche had swept away a weather-recording station at Alta, Utah, in 1951. Firstly, it was soon obvious that the principle of searching the fall-line below the disappearance point was valid. And Atwater also noticed a definite tendency for the debris to be grouped in zones, although within the zones there was no pattern for depth of burial according to the weight or shape of the object. This observation is useful in confirming that it is correct immediately to search the area close to the point at which any of the victim's clothing or equipment is found on the surface.

The grouping of debris is doubtless due to the wave action of avalanches and to the effects of terrain, factors which are themselves interrelated. On film, especially, it is quite common to see the front of an avalanche slow down and a wave of snow from behind break over it and assume the lead. This is often caused when a feature of terrain slows the front of the avalanche. It is at the point where a wave breaks over another that debris may well be deposited, so a good Accident Site Commander tries at once to establish whether such a wave pattern exists, and at what points a feature of terrain may have slowed the front of the avalanche and produced a dropping zone.

A summer knowledge of the slope in question can be invaluable in helping him to assess the situation, as was the case in a rescue carried out by the Parsenndienst a few years ago. The Accident Site Commander began the search very high on the slope and, at the time, there was much adverse comment on his tactics. But almost immediately the search proved successful. What the Accident Site Commander had known, and his critics had not, was that he was searching at a point where a path ran across the slope in summer. After the passage of the avalanche the slight terrace effect of the path had been obliterated, but it had been sufficient to slow the front of the avalanche and cause the victim's body to be deposited.

The uphill side of a hummock over which the avalanche has run is the commonest dropping zone and should claim the attention of the Accident Site Commander at once. But even when he has decided on priority areas, his troubles are not over. It can be that all the evidence points so forcibly to a certain priority area that he cannot accept the 30% chance of missing the victim if he orders the area coarse sounded. But if he orders it fine sounded, and the evidence has been misleading, the loss of time could well allow the victim to die, when coarse sounding a larger area might have saved him.

If after several hours of coarse sounding the victim has not been found, fine sounding is usually begun; by then the recovery of the body rather than a live rescue has become the chief objective. If they work well there is very little chance of the sounding team missing the body when fine sounding. They must push the rods down vertically and they must know what they are feeling for—that is to say that resilience under the rod that no rock, earth or tree-trunk can impart. They should always wear gloves otherwise the warmth from their hands causes snow to freeze to the rod. However, debris like rocks and trees in the avalanche can reduce their chances of locating the body, as can the passage of time; after several days a body can be frozen so hard that it feels like the ground when struck with a rod. If sounding rods fail to locate the body the only alternatives are trenching through the avalanche with shovels, or the unpleasant one of waiting for the spring thaw.

*   *   *

Let us return once more to the difficult task of finding an avalanche victim who is still alive. Sounding-rods are essentially too slow and the best alternative today is the avalanche dog. These highly trained and intelligent animals have saved many lives in recent years. *A dog can search an area of* 100 *by* 100 *metres in* 20–30 *minutes, against the four hours taken by a team of* 20 *men coarse sounding. A really exhaustive search of the same area with dogs would only take two hours, against the* 20 *taken by the men fine sounding; but, because few dogs will work at their best for more than half an hour at a time, this presupposes that at least two dogs work in shifts.* This performance is so remarkable that it is worth devoting some space to avalanche dogs, to their origins, training, and use today.

The forerunners of the modern avalanche dog were those vigorous animals kept at the St. Bernard Hospice, where they can be traced back to the latter half of the 17th century. The St. Bernard type of dog has been known much longer, however, and is probably descended from the Tibetan mastiff brought to Greece by Xerxes about 485 B.C. From there it was brought further west in Europe by the Romans, and by the 14th century dogs of the type were beginning to appear on coats of arms. The St. Bernard Hospice recruited their first dogs to guard them from brigands and highwaymen, a task which they performed effectively. According to one story, a group of highwaymen once partook liberally of the Hospice's food and wine and then, as they were about to leave, demanded that the strongbox be given to them as well. The Prieur said he would fetch it but came back with a pack of bristling dogs instead.

The highwaymen relinquished their claim to the strongbox and left in some haste.

Later the guides began to take dogs with them on their daily sorties from the Hospice, both for company and for their ability to cut a track through deep snow. In fact, this latter was considered a very important task; with their unerring sense of direction they would bound powerfully ahead, leading the guide and his party safely through the deep snow and blinding storms. It is an interesting point, incidentally, that the true and original St. Bernard dog had a short coat, unlike the present dogs bred in England and America whose shaggy coats would gather too much snow and make them unsuitable for work in the mountains.

Much romantic nonsense has been written about St. Bernard dogs; this is a pity because the unadorned truth is interesting enough. In the first place the dogs began their rescue activities spontaneously. When a guide was out looking for people in fog or storm the dogs soon realized the purpose of the exercise and, once they had picked up the scent of the missing traveller, they would lead their master to him. It was a small step from this to smelling out people who had fallen from exhaustion and been lightly covered with snow, and from that to locating people shallow-buried in an avalanche. It is doubtful whether they had the almost miraculous powers of the specially trained avalanche dogs of today, but it is known that they saved several lives by scratching at the snow over an avalanche victim and so disclosing his whereabouts to the monks.

The most famous of the dogs was Barry I who, in the first 12 years of the 19th century, saved more than 40 lives. One one occasion, Barry found a child lying in the snow. He licked the child's face until it regained consciousness and it was then able to cling to the dog and be half carried, half dragged to the Hospice.

Barry I, who became world famous, died in 1814 after spending two leisurely years of old age in the lowlands. His stuffed body is now in Berne Museum and his name, which comes from the German dialect *Bari* meaning 'little bear', is perpetuated at the Hospice by giving it to the best male dog in the kennels.

Two popular beliefs about St. Bernard dogs should be discredited. Firstly, though Barry I went out alone on many occasions, it is not generally true that the dogs worked independently; they carried out their rescues with members of the Hospice staff.

Secondly, the carrying of the little barrel around the neck is also the figment of someone's imagination, perhaps that of the erstwhile publicity manager of a certain brandy manufacturer. The basis for the belief may have derived from the fact that one or two dogs were trained to wear a small saddle with two containers for carrying milk from the cowshed to the Hospice.

Strangely, it was not the St. Bernards that led to the modern avalanche dog but an event which took place in the winter of 1937/38. A group of young people on a ski-tour were buried in an avalanche on the Schilthorn above Mürren. One of the rescue team had with him his little terrier-like dog called Moritzli. All the victims but one had been found, and the rescuers were still searching, when someone noticed that Moritzli kept returning to one place. The area had already been sounded so no one attached any significance to the fact—until Moritzli began to bark and whine. The rescuers re-sounded the place and found the young skier, still alive.

The Moritzli incident took place at a fortunate time; the mountains were being frequented by ever more people and war was on its way. When a dog expert called Ferdinand Schmutz, of Berne, heard about Moritzli's performance he began propounding the idea that perhaps dogs would be *trained* to find avalanche victims. The Swiss Army was delighted and asked that attempts be made at once. Four Alsatians (German Shepherd dogs) were duly trained and presented to the Army.

After the war the Swiss Alpine Club became interested in the possibility of having avalanche dogs available for their own rescue network, and they took on the responsibility of organizing training courses. Later, various dog clubs in the Alps also entered into the activity and the training of avalanche dogs has now become so widespread that courses are run in Switzerland, Austria, Sweden, Norway, Germany, Italy and Yugoslavia. In Scandinavia the Red Cross sponsors the courses.

For a number of years after the war, no dog saved the life of an avalanche victim, but there was a simple reason for this lack of success: the dogs were so few and far between that there was scant chance of having one at the avalanche before the victim had died. In more recent years, however, they have saved scores of lives and located even more dead bodies. Of late the chances of success for a dog have been increased by air-lifting them to the avalanche. In the early days of the *Rettungsflugwacht* attempts were made to parachute the handler and his dog on to the avalanche, but transporting them by helicopter has been found more suitable.

In about half the rescues in which dogs take part today they give a positive result, that is to say they find the victim, either dead or alive. There is less chance of a positive result if the snow of the avalanche is very wet or compact, and in these conditions it is generally accepted that a dog is unlikely to find a body buried deeper than 4 feet. In dry snow there is virtually no limit to the depth at which a dog can find someone, if other factors are favourable. Among these other factors is the air temperature, because very low temperatures definitely have a detrimental effect on the dog's ability. But they are more tolerant than one might suppose and have been successful down to air temperatures of −32°C (−25°F). Normally, however, it is held that −20°C (−4°F) is the lower limit at which a dog is likely to succeed.

Whether the victim is alive or dead is also important: more scent is produced by a living body, or by a warm one, than by a cold corpse, at least until putrefaction sets in. *However, it is generally the overall conduct of the rescue and the handling of the dog by its master that are the instrumental factors in determining the dog's likelihood of success.* When dogs have failed, a subsequent enquiry has frequently shown that the blame can be laid upon the humans at the rescue.

The basic requirements in a dog if it is to be trained for avalanche rescue are intelligence, a willing temperament and sufficient size and strength to cover large distances through deep snow before it even begins to work. To date by far the most successful breed has been the Alsatian, in any case renowned for its intelligence and ease of training. The Groenendaal (Belgian Shepherd) has been fairly successful although it is more erratic and temperamental than the Alsatian. The odd St. Bernard and Samoyed have also been trained.

In Switzerland, training courses are run every autumn, usually in November, and anyone with a suitable dog can attend if he wishes. There is, though, a prejudice against bitches. This is because it is not unusual for several dogs to assist at one rescue and if one of them happens to be a bitch on heat, the others are put off their stroke.

Dogs usually begin their training at about a year old, but dog and master must attend three courses before they can qualify for Class C, the highest standard of proficiency. The training itself is not complicated but, of course, requires patience. The first step is for the dog's master to step down into an open hole in the snow while the dog watches. He then calls the dog to him and rewards it well. This is repeated without calling the dog, and then the dog is led out of sight while its master gets into the hole.

When the dog has understood this game, and runs unfailingly to its master as soon as it is released, a thin covering of snow is shovelled on to the man while the dog watches. And the next time, the dog is taken away during its master's burial. Sometimes, when the dog is released and reaches the edge of the hole it has a moment of obvious perplexity, but then it catches its master's scent wafting up through the snow and starts to scratch for him. The next stage is to place a stranger in the pit, with the master lying under him, and to cover them both with snow. The dog is especially rewarded when it finds the stranger. After this the switch is made and the master accompanies his dog while it searches for a stranger.

Once the dog has reached this stage satisfactorily it qualifies for Class A. Experience shows that if handled properly and rewarded sufficiently only about 10% of dogs fail to complete this basic training.

During its year in Class A a dog receives further practice with its master, and it may be called to a real rescue if no experienced dog is available. During its second training course, attention is given to the dog's method of searching. Having been allowed free range to start with, it is gradually persuaded to zig-zag, thus learning the beginnings of a systematic coverage of the avalanche. During this second course the dog is expected to find one man, then two men, then two men and two objects. The object is usually something like a rucksack, but in good conditions it is incredible what a dog can find. They have been known to locate ski-sticks buried 3 feet deep, and the only part of a steel ski-stick likely to give off scent is the leather handle. With its second course completed satisfactorily the dog enters Class B.

The next year, at its third course, the dog perfects a systematic search pattern for covering an avalanche, and is expected to find men and objects buried much deeper. If the dog performs well it will be promoted at the end of the course to the elite Class C. To maintain its status thereafter the dog must attend a refresher course every second year.

During the courses almost as much emphasis is placed on the training of the master as on that of the dog. The master must be fully conversant with the ways of avalanches and with general rescue procedures if he is to direct his dog to the best advantage. He must be ready to insist that his dog be allowed to work in the most favourable possible conditions. A poor Accident Site Commander may have allowed equipment to be dumped on the upwind edge of the avalanche, and he may even fail to clear men away from that side when the dog arrives. If he has been inefficient in this way, the master must demand that the situation be rectified before he sends in his dog. In the event that the sounding team has not been withdrawn before the dog's arrival, a few minutes must be allowed to pass with no one on the avalanche so that any human scent can clear before the dog begins its search. In any case, the dog will benefit from a short rest if the journey to the avalanche has been arduous.

It is all too easy, in the heat of the moment, for an owner to put his dog to work regardless of circumstances, especially after a helter-skelter dash to the scene. Indeed, he will have to show great strength of character if he is to insist on measures which cause delays, even though the measures taken may bring about the dog's success and save hours of searching in the end. An owner must also have the courage to call off his dog for a rest every 20–30 minutes, for few dogs work well for long periods at a time.

An incident occurred on a dog training course a few years ago which, apart from being somewhat amusing, also illustrates the need for coolness on the part of dog handlers. The course was being run at Davos and the training was being carried out at the bottom of a slope. Owing to a change in weather, the Parsenndienst decided one morning that the slope might be dangerous and they therefore asked the leaders of the course to clear the area until a certain time so that they could try to release an avalanche intentionally.

When the patrolmen looked down from the ridge at the head of the slope, however, they were surprised to see that the men and dogs were at its foot as usual. The patrol leader was about to shout to them to move away when something, perhaps the disturbance of his skis on the ridge, released the avalanche. The course saw it coming and only had a short distance to run to escape its range. But the head of the avalanche covered the place where a man had already been buried in the snow as part of an exercise.

Such unbridled pandemonium broke out that the patrolmen on the ridge could but laugh. Everyone began to shout, men poked here and there with ski-sticks, and dogs ran about in all directions before order was restored. The dog belonging to the buried man was rushed in to search but, after very few minutes, it was called off, and the probing with ski-sticks began again. With the aid of the Parsenndienst the man was eventually found after a burial time approaching 30 minutes. He was unhurt although very unnerved by the experience. Even in a normal exercise it is not very pleasant to be buried in snow. (It is usual to give the volunteer a torch and a good book to take his mind off the possibility of not being found, as well as a cylinder of oxygen if he is to be buried very deeply.) And suddenly to know from the extra weight that an avalanche has added to your burial depth must be extremely worrying.

It is easy to be wise in retrospect but the correct action on that occasion would have been for the whole course to do nothing for 7–8 minutes. During that time the man's scent would have percolated through the additional snow of the avalanche and one of the dogs could then have found him. The man was in no immediate danger and the delay of a few minutes would have resulted in a quicker rescue overall.

A truly remarkable incident involving avalanche dogs occurred near the Mauvoisin dam in the Val des Bagnes, Valais, in 1952. On many of the large dam-building schemes in the Alps there were, and still are, safety and rescue teams—similar to the Parsenndienst but much smaller—for the protection of the construction company's personnel and installations. These teams were often led by mountain guides, and Louis Wuilloud was in charge of the team at Mauvoisin.

In the early evening of November 10th, 1952, two workers set out down the road from Mauvoisin towards Fionnay, choosing to ignore the fact that all use of the road had been strictly forbidden by Wuilloud because of acute avalanche danger. They had not gone very far when they heard an avalanche; one of them saved himself by leaping into the mouth of a gallery, but his friend was not quick enough, and the avalanche buried him. Louis Wuilloud's rescue team and an avalanche dog went straight to the place when the survivor raised the alarm; but in many places the avalanche was 30 feet deep, and, having found nothing, the team withdrew because of the threat of further avalanches.

Some Swiss border guards in the area, who owned avalanche dogs, had also been informed of the accident, and they at once set out to render assistance. Two guards, Spalinger and Monnet with their dogs Ello and Astor, and a third guard called Meli, reached Fionnay at 03.30 after a difficult journey from Vintsay. Spalinger and Meli had gone to Vintsay to meet Monnet, who had come from Bourg St. Pierre by jeep, but the road between Vintsay and Fionnay was blocked by an enormous avalanche.

At Fionnay, they were asked not to set out for the scene of the accident until daylight, so they left just after 07.00. Their journey up the road towards Mauvoisin was extremely dangerous. A number of avalanches had come down, and the guards kept a spacing of 50–60 yards between them. (They could not spread out more without losing sight of each other on the twisty road with its piles of avalanche snow.)

At about 07.40, a large loose-snow avalanche came down on them. It was at least 200 yards wide and it buried all three men and both dogs, but luckily not very deeply. The dogs were able to free themselves. Spalinger was trying to fight his way out of the snow when they both arrived and dug frantically to help him. No sooner was he free than Astor, Monnet's dog, galloped off towards the bed of the River Dranse. Spalinger followed, and the dog led him straight to Monnet, about 120 yards away. Monnet had been able to free his head with Astor's help, but apart from that he could not move at all.

The avalanche had dammed the River Dranse, and Monnet was surrounded by a water and snow mixture that was still rising. Spalinger and the two dogs dug and struggled for about 10 minutes to get him out, but he was held firmly by his skis. Finally, by lying down and plunging his arm and shoulder into the water mass, Spalinger reached Monnet's ski-bindings and released them.

Of Meli, the third guard, there was no sign; but the dogs were soon quartering backwards and forwards, and they located him. He was buried at a depth of 18 inches, but he was also dug free alive. The initiative of Astor and Ello doubtless saved the situation. Monnet said, afterwards, that Astor had come to him immediately after the avalanche and had helped to free his head; then, without any command but probably because he knew he needed human assistance, he had gone off to find Spalinger and help him free.

For the two days after their accident, the three guards and two dogs helped from dawn to dusk with the search for the body of the worker. But the avalanche debris soon turned to ice, as a result of the damming of the Dranse, and the search had to be abandoned. The body was found in the spring thaw, six months later.

*   *   *

As we shall see towards the end of this chapter, recent years have seen progress in the development of some interesting electronic devices for avalanche rescue, but there are still snags in their general applicability. And so sounding rods and dogs are still, at the time of writing, the commonest means employed in rescue operations; and the only way to increase the likelihood of successful rescues with these means is through more speed and efficiency. But this need for efficiency does not only apply to the rescue team because, in almost every case where there has been a happy outcome to an avalanche accident, the foundations for a successful rescue have been laid by the unburied friends of the victim, or by a casual observer of the accident. Cool action during those first few minutes is of great importance, and if more people knew what to do, more avalanche victims could be saved.

With a glance to make sure that no more snow is likely to come down, any witnesses should hurry on to the avalanche and mark, with ski-sticks, the point from which the victim was taken and his disappearance point. People are often afraid to go on to a newly-fallen avalanche but it is nearly always safe to do so unless it is in a gully. In that case it is possible for a secondary avalanche to fall on to the first, but even then the risk is not great because the disturbance of the first avalanche has usually dislodged any others that are ripe.

The witnesses must then make a rapid superficial search of the avalanche to see whether there are any signs of the victim like an item of equipment, the tip of a ski, or the fingers of a hand protruding from the snow. If there are no clues the fastest skier should be despatched at once for help. (If help is a long way off it may be wiser to send two people.) The remainder of the party stays behind and begins sounding with ski-sticks, skis or Lindenmann rods if they are available.

Some quite unnecessary tragedies have taken place because these rules of procedure were broken. For example, before the last war three members of a ski-party were buried in an avalanche while on a tour. Those not buried rushed for help immediately, without even staying for the most cursory look around. When the rescue team arrived several hours later, after a long climb, they found that a ski-tip of one of the victims was actually poking out of the snow. When they released the body, which was only buried a few inches deep, they discovered that the unfortunate man had cleared a large space around himself in a desperate struggle to get free; but he had died by the time they reached him. Even when there is but a solitary witness to the accident he should make a quick superficial search before going for help.

People have been known to react to an avalanche accident in the most disastrous way owing to ignorance and panic. As an example of this, nothing can quite cap the incident at Zermatt in March 1962, when Norbert Julen, a 27-year-old ski-instructor, was giving a private lesson to a woman on the Hohtälli run. Julen was leading, slightly off the marked track, when he released a tiny snow slide. Its fracture line was a mere 6 inches high and it only ran 30 yards, but even this insignificant affair was enough to throw Julen off balance and pitch him headfirst into a small hollow.

His head, shoulders and part of his torso were buried but the lower half of his body was free. His pupil, instead of pulling him out with a sharp tug on his legs, or digging his head free with her hands, skied off for help. The rescuers arrived after 30 minutes, but by then no amount of artificial respiration could revive Julen. Surprised as he was by the slide, he had allowed his face to be rammed into the snow so that only prompt help could have saved him from suffocation.

In the Parsenndienst headquarters a klaxon alerts the staff when an avalanche accident has occurred. The raucous bray introduces a period of desperate activity, but nevertheless, a period devoid of panic. The initial step is the despatch of the First Rescue Team under the leadership of the most experienced patrolman available—experience is important, of course, because he will become the Acci-

24. Georg Caviezel throwing a *Sprengbüchse*

25. The rescue of Caviezel from the avalanche that swept him away shortly after photograph 24 was taken

26. Avalanche damage in an area of new houses at Davos in January 1968

dent Site Commander on arrival at the avalanche. He and his men are on their way within five minutes of the alarm being raised.

They take with them what is called *Avalanche Sledge A*. This sledge, which is always kept ready, is loaded with 30 non-collapsible sounding-rods, avalanche shovels and coloured flags for marking important features at the avalanche. (As long as they can be transported downhill on a sledge, non-collapsible rods are perfectly satisfactory and they have the advantage that the inexperienced volunteers, who join in the rescue later, find them easier and quicker to put to use than a rod that needs assembling.)

The Leader of the First Team carries the megaphone he will use at the accident plus all the other items he will need as Site Commander. Someone else in the Team is equipped with radio. If the accident had occurred during a period of acute avalanche danger, it is likely that a dog and its master will have been stationed in the headquarters, and in that case they too would join the First Team.

Dependent on the manpower available for the First Rescue Team they may, or may not, take with them *Avalanche Sledge B*. This sledge is loaded with artificial respiration pumps, pumps for sucking moisture and snow from a victim's throat and mouth, an oxygen cylinder, blood plasma, heart injections, tubing to be passed into a victim's lungs, surgical instruments so that a doctor can perform a tracheotomy, tea, hot-water bottles and blankets. If the First Team cannot take this sledge with them it will follow a few minutes later with the first of the Reinforcement Columns.

The five-minute period before the First Team leaves could perhaps be reduced, but it is better to take a little longer and leave properly prepared, especially from the viewpoint of precise orders. The Leader must know exactly where the avalanche has taken place, for were he to overshoot, the climb back or despatch of a second team would involve a drastic time-loss. In fact, without precise orders a rescue can turn into a second accident. During a rescue in the United States in 1958 two uncoordinated parties set off for the same avalanche. One party, which followed high ground, released an avalanche on to the party below and one of the rescuers was killed. Indeed, the thought which must always remain in the minds of the rescuers is that they are threatened by avalanches too, for if there were no danger the accident would not have occurred in the first place.

While the First Team is rushing to the avalanche the Rescue Manager has in use every telephone line at his command. He calls for avalanche dogs, doctors and Reinforcement Columns which will include volunteers requested over the loudspeaker in the Weiss-fluhjoch restaurant. He also calls in the help of the Avalanche Research Institute whose men are immediately available to support those of the Parsenndienst. He informs certain railways and cable-cars of the accident and asks that they do their utmost to ensure the rapid transportation of vital personnel from the valley. In an area as well developed as the Parsenn it is unlikely that the *Rettungsflugwacht* will be needed, but had the accident occurred in a remote district they too would be called by the Rescue Manager in those first minutes.

On arrival at the avalanche, the men of the First Team leave their sledge(s) at the top and ski down the avalanche track looking for any signs of the victim. They mark the edge of the avalanche with yellow flags so that no confusion can arise when the whole area has been trampled on. They mark with crossed flags any significant points like the skier's entry track. The Accident Site Commander quickly obtains information from any witnesses, makes his vital decisions and sets the men to work with sounding-rods. By means of the radio, headquarters are kept in the picture, and in return the Accident Site Commander is informed about the likely arrival of reinforcements so that he can plan accordingly.

The sounding goes on at a feverish pace—in silence. Only the Accident Site Commander is allowed to talk, not merely because this increases the faint possibility of a victim making himself heard, but also because someone might pass some remark to the effect that he did not think much of the victim's chances. If the victim were to hear such a remark it would do untold damage to his morale, whereas the sound of activity and crisp orders alone bring immense comfort and hope.

As areas are searched they are marked off with red flags, close together or spaced out to distinguish between coarse and fine sounding. When the Accident Site Commander estimates, from information received by radio, that the avalanche dog is within some 10 minutes of arriving, he calls off the sounding team and moves them, and all equipment, downwind of the avalanche. He has previously told them not to drop anything like a cigarette-end on the avalanche and only to urinate well away from the area.

Contrary to some beliefs, it is a dire mistake to forbid the avalanche to sounding teams until a dog has searched it, because a 10-minute period with no one on it is quite enough to clear the air. The Swiss Army once made this grave error when two soldiers were buried in a tiny avalanche. The rescuers stood back with their rods and shovels when they could have searched it completely before the dog's arrival. The dog located the soldiers right enough, but by then one had already died.

The dog and master go on to the avalanche together and begin the search. If they are unsuccessful, any other dogs that may have arrived will be given a chance before sounding is resumed. After several hours, the initial rescuers will be relieved of their exhausting work by fresh teams from the valley, well fed before they set out. With the approach of night a petrol-engined lighting plant is brought to the scene and the work goes on without interruption.

When a victim is located by sounding-rod, or dog, his head is dug free as quickly as possible and revival attempts begin while the rest of his body is being cleared. Mouth and breathing passages are freed of snow, mucous, vomit or blood with the little foot operated suction pump (seen by the patrolman's knee in photograph 19c). Then artificial respiration begins, either by the mouth-to-mouth method or with the *Ambu* bladder pump and face mask equipment (also seen in photograph 19c). The advantage of the pump is that it can be used for long periods without fatigue, and there is provision for feeding additional oxygen into it. Injections of heart stimulant and heart massage accompany artificial respiration. By this time there is usually a doctor on the scene and he will supervise the work and make any surgical incisions necessary. In bad weather the victim is often moved into a tent while the attempts to save his life continue. If the attempts are fruitful, the victim is loaded on to a sledge as soon as the doctor considers him fit to withstand the journey to hospital.

A search is never abandoned while there remains the remotest hope; but the safety of the rescuers has to be considered and it may be necessary to halt the action for a while so that other threatening avalanches can be released. (During an infamous rescue in 1951, six men were killed by second and third avalanches down the same track while they were trying to find the single victim of the first. Understandably this, and similar examples, continue to haunt rescue managers.)

Even when it is virtually certain that the victim is dead, the search continues for his body, although ultimately it may become economically unjustifiable to go on. Only once has the Parsenndienst reached that point and abandoned a body for the spring thaw to reveal its whereabouts. A young man had released an enormous avalanche on the Dorfberg just above Davos. After many days of sounding and of searching with dogs without result Christian Jost called off the official rescue, but Hans Kerschbaum, one of the two Reserve Leaders of the Parsenndienst, continued to work over the avalanche time and again with any available staff.

The young man's mother then arrived in Davos and complained that, because she was not rich, no effort was being made to find her boy's body. The matter preyed more and more on Kerschbaum's mind until at last, some six weeks after the avalanche and after a period of warm weather, he found the tip of a ski. The body was retrieved from under a ledge of rock which had been screening it from sounding-rods.

Just occasionally an avalanche rescue takes place which is not only carried out with great speed and precision, but is also favoured by that uncommon degree of chance and good fortune without which no amount of efficiency alone can lead to success. The Parsenndienst brought off such a rescue on March 16th, 1958, and it is particularly interesting because every actor in the drama was not only available later to describe his experiences, but was also articulate. It was a copybook rescue but with several odd factors helping it to succeed.

Indirectly, the success of the rescue was perhaps determined some two days before the accident. On March 14th a German doctor, Marcel Kurz, and his wife, who were attending a medical congress in Davos and combining some skiing with business, were travelling to the Weissfluhjoch in the Parsennbahn. They overheard and then joined in a conversation which was to have unimagined significance two days later.

One of the railway staff was talking about avalanches with a large man in Parsenndienst uniform. Dr. Kurz and his wife discovered that the man was Christian Jost, but they did not discover that the conversation had probably arisen from the fact that two skiers had been killed by avalanches in the area during the preceding eight days. When they joined in the conversation the topic of Jost's own avalanche burials came up. With his usual modesty Jost merely remarked that he had remained near the surface by swimming; and he then dismissed the subject by saying that, in any case, a patrolman's colleagues always had him out in an instant and that the victim received a 20-Franc bonus for his pains. Jost was, of course, playing down the matter rather than risk being a sensation-monger.

The fortunate effect of this was that Dr. Kurz and his wife, who were almost totally ignorant about avalanches, were left with the impression that they were an overrated menace. After all, reasoned Dr. Kurz, it could hardly be an acutely dangerous experience that only drew 20 Francs compensation. (The bonus has since risen, in line with inflation.) In addition, they were greatly impressed by Jost and filled with faith in the ability of the Parsenndiest to rescue avalanche victims if the worst came to the worst.

Sunday, March 16th, was to have been the last day in Davos for Dr. Kurz and his wife, and they were delighted when the snow-storm of the previous days gave way to a fine morning. When they reached the Weissfluhjoch the signs were out warning skiers of avalanche danger and telling them to stay on marked runs.

They decided on the Meierhof as their last ski-descent before getting into their car and driving to Germany. They set out correctly, but part way down they branched away from the poles marking the run and entered the Meierhof gully. There was not a track in the deep snow that was becoming heavy under the effect of the strong sun beating into the hollow. At about 11.40 the snow was so sticky that they stopped, right near the exit of the gully, to wax their skis for the remaining few hundred yards of the run.

It was one of those blazing Alpine days of such serene limpidity that one's very soul seems purified. There was nobody in sight; the sun was hot and, in no hurry to move on, Dr. Kurz and his wife removed their anoraks and continued peacefully to wax their skis. Suddenly the solitude was broken by the appearance of two skiers at the head of a slope above them. At first Dr. Kurz was irritated by what seemed almost an intrusion and then the thought came to him that the skiers might release an avalanche on to them.

Dr. Kurz was on the point of calling to them to be careful when there was an ominous click, like the snapping of fingers, somewhere on the slope behind him, the slope opposite to the one on which the skiers were. An instant later the urgent shout: '*Achtung! Lawine!*' rang out from the strangers.

'Run!' said Dr. Kurz, and without even looking over their shoulders he and his wife floundered off down the gully. The evil hissing noise behind them swelled in their ears and they were suddenly hurled forward on their faces. They flung up their arms as they fell. A great weight swept over them and enveloped them; and then, as the movement slowed, they could not breathe. Finally the avalanche stopped and the pressure on their bodies eased somewhat.

Dr. Kurz's first thought, as he was pitched forward, was of the irony of being buried by an avalanche within minutes of the end of his skiing for the year. Then, when breathing was difficult, he remembered a patient choked by an oedema in his own sitting room; he remembered how he had cut open the man's trachea and he wished that there was someone to help him in his need. Once able to breathe again his thoughts were mainly of anger at his fate and worry for his children, but he had no fear as such. He believed later that this was due, partly at least, to the conversation with Jost two days earlier. He also remarked, with a smile when talking to me, that death held no terrors for him since the Second World War during which the Allies had frequently presented him with its imminent probability. As he lay under the avalanche he shouted once for help, but the noise he made seemed laughably trivial. After what he estimates as a minute he lost consciousness.

Frau Kurz also thought about her children. She wondered how far away her husband was, and then her thoughts turned towards the Parsenndienst. She imagined what steps were being taken to rescue them, and it never occurred to her that they might not be saved. She believes she lost consciousness in about three minutes.

The two skiers who had shouted the warning watched the avalanche overtake Dr. and Frau Kurz. One of them glanced at his watch and noted that it was 11.50. As soon as the avalanche had stopped they made their way on to the tumbled snow, marked the disappearance point of the victims, and then, while one of them began a superficial search, the other went for help.

At 12.02 the telephone in the Parsenndienst headquarters rang; an instant later the klaxon sounded and the rescue action had begun. All the various calls for avalanche dogs, Reinforcement Columns, doctors, etc., were made and the First Rescue Team set out with the avalanche sledges.

At the avalanche, the witness who had remained was soon joined by three other skiers and together they completed the superficial search. They found a ski-stick, a camera, a pullover, an anorak and a pair of gloves, and they began to sound with inverted ski-sticks in the area where these items had been deposited. At 12.25 the First Rescue Team of the Parsenndienst arrived and took over the operation. Meanwhile the reinforcements were gathering fast.

Melchior Schild, a member of the staff at the Federal Institute for Snow and Avalanche Research, and one of the great experts on avalanche dogs, was quietly beginning his Sunday lunch at home when the telephone rang. A moment later he and Iso, his fine dog, were on their way, all thoughts of food forgotten. Iso, who appears with his master in photographs 19a and b, was one of the most successful of all avalanche dogs. By the Austrian points system of 3 per search action, 15 per dead body located, and 30 per life saved, Iso would have accumulated 111 points in his lifetime. So keen was he that he would howl with excitement when he realized that there was work to be done. But he had a hard journey ahead of him before reaching the avalanche containing Dr. and Frau Kurz.

Schild was in a quandary as to the quickest way to reach the avalanche: either to take the Parsennbahn to the Weissfluhjoch and ski down, or to go along the valley by car and climb up the Meierhof gully. He decided on the former course and Iso was set the task of covering, through deep snow, a distance of 2½ miles with a height drop of 3,000 feet. He did it in 20 minutes.

At the avalanche another problem beset Melchior Schild: should he make Iso search for both victims immediately, or for one at a time? Bearing in mind the delay while the first was dug out if Iso found them separately, he decided to try for both at once. The avalanche field was clear; Iso had rested for a moment, and everyone was standing downwind in silent expectancy. With the urgent command: 'Search, Iso! Search the man!' Schild released the dog.

The Alsatian set out with a will, quartering to and fro. After three minutes he stopped and somewhat half-heartedly began to scratch. Schild made a careful note of the place and called Iso to him before sending him off again. The dog went a little further to the right and after five minutes suddenly pulled up in mid-bound, sprang to one side and began to dig frantically. Waving aside the anxious rescuers with their shovels, Schild called Iso to him yet again and sent him off in the general direction of the first spot. Sure enough, the dog went straight to it and scratched more urgently than the first time. Then he ran to the second spot and scratched again, looking round at his master as if to say: 'Here, you fool!'

Schild dug hurriedly and came across a man's shoulders buried just over 3 feet deep. The shoulders moved. Digging in the other place indicated by Iso he found a boot and ankle, some 4 feet deep.

At 13.40, that is to say 1 hour and 50 minutes after their burial Dr. and Frau Kurz were lifted clear of the snow. Both of them were purple in the face but the outlook, for Dr. Kurz at least, appeared favourable. He wavered between consciousness and unconsciousness for some time, his eyes occasionally flickering open, but under treatment with oxygen and heart stimulants he was soon regaining strength.

For his attractive, fair-haired wife, however, the outcome remained in doubt for several hours. When her head was first freed her jaw was clamped so firmly shut that it was impossible to pass the small tube of the suction pump into her mouth. In a flash Hans Kerschbaum smashed off an eye tooth with the handle of his pocket knife and inserted the tube.

There were four doctors present, all of whom had been attending the congress with Dr. Kurz and who, by the greatest fortune, were in the Weissfluhjoch restaurant when the call for volunteers went out. Between them they mounted a desperate battle for Frau Kurz's life. It was almost two hours before she was pronounced fit for the journey to the valley, but even then she was still unconscious. A few minutes later she was in a room at the Kulm Hotel in Wolfgang where she received further treatment and massage to help her circulation. From the hotel she and her husband were taken to the nearby sanatorium in an ambulance, but her state was so precarious that she was twice given artificial respiration on the way.

Once in the sanatorium her condition began to improve. Dr. Kurz was allowed to leave and his wife was transferred to Davos Hospital for two days. Four days after the avalanche they drove home to Stuttgart.

Dr. Kurz suffered no marked effects from shock until about ten days after the accident when a relative brought in a box of colour transparencies that he had had developed. The film was from the camera that had been found on the surface of the avalanche and, unbeknown to Dr. Kurz, someone had taken a photograph of the two holes in the snow from which he and his wife had been saved. When he saw the photograph unexpectedly the shock was so great that he was forced to lie down.

When I interviewed Dr. and Frau Kurz, they talked about their experience in a completely matter-of-fact manner, but at the same time there was an evident sense of wonder and gratitude that they escaped with their lives. They remarked on the fortunate set of circumstances, like the conversation of two days earlier which gave them so much faith in their rescuers, a faith which was certainly an important factor in their survival. They remarked on the fortuitous presence of the two skiers who saw the accident and acted correctly; and they remarked, too, on the fortunate fact that they had removed their anoraks and pullovers. Dr. Kurz believed that this contributed to their survival by allowing their bodies to become undercooled more rapidly, thereby reducing their oxygen needs.

They were grateful to the men who spared no effort on that fateful day. And above all were they grateful to Iso, a dog whom they later came to know well, and to love—a dog whose exertions in the service of man killed him at the early age of nine.

*   *   *

In recent years, much time, effort and money have gone into trying to develop better means of locating avalanche victims than by sounding rods and a dog's fine sense of smell. Dogs are, in fact, very effective, but usually they arrive at an avalanche too late to find the victim(s) still alive. And as we have seen, sounding rods are painfully slow to use.

An institution which has had a central role in much of the research and development work of recent years is the International 'Vanni Eigenmann' Foundation of Milan, Italy. Vanni Eigenmann, of Swiss extraction and a keen skier, was a member of a prominent Milanese family that has large chemical engineering interests. He himself was a brilliant engineer who had studied in the United States and Britain and who was being groomed to take over the family concern. But in February 1961, at the age of 35, he ventured into the notorious Val Selin near St. Moritz and released an avalanche that ran almost a mile and buried him. (The Olympic skiers Bud Werner and Barbi Henneberger were killed in the same gully while filming in 1964.)

The search for Vanni Eigenmann turned into the longest avalanche rescue in history. Sounding teams worked up and down the enormous avalanche; many dogs were tried; sorcerers and *clairvoyants* made incorrect predictions as to where the body lay; and shovel gangs were imported from Italy. Finally, on the 37th day, a German-built Förster magnetic detector, or magnetometer, located the body by means of the steel parts of the ski-boots. Vanni Eigenmann's uncle, Dr. Gino Eigenmann, was the prime mover in this protracted search. It was, of course, very costly and local people poured scorn on such a waste of resources. After all, they reckoned, spring would sooner or later show where the body lay.

Gino Eigenmann also alienated a lot of people in St. Moritz by his justified criticism of the sloppy way the rescue operation had been handled in the first hours after the avalanche, when Vanni might still have been alive. However, people were forced to change their views about the Eigenmann family when, quite soon after Vanni's death, they announced the creation of the International 'Vanni Eigenmann' Foundation with the purpose of sponsoring and financing the development of effective devices for avalanche rescue.

Apart from promoting and financing research and development, the 'Vanni Eigenmann' Foundation has organized two symposia to bring together mountain rescue and electronic specialists, the first in Davos in 1962 and the second in Solda, Italy, in 1975.

In the early days of its activities, the Foundation concentrated its efforts on attempting to develop a device that would locate a body through its intrinsic qualities. Some of the principles considered, tried, and mainly rejected, were: detection of the CO_2 exhaled by the victim; ultrasensitive gravimeters that would reveal the presence of a body by differentiating between its density and that of the surrounding snow; locating a body through its radiation of warmth; radar; and listening for the heartbeat or breathing of victims.

The purely technical difficulties of developing a device capable of locating a body buried in snow are daunting, but when one adds to them the general prerequisites for an avalanche rescue device, they become even more formidable. It must be fast, allowing a coarse search of an area 100 metres by 100 metres in a maximum of 30 minutes—the same time as for a dog; it must be easily portable, simple to use, robust enough to stand inevitable shocks and the effects of mountain weather; and it should not cost too much. Nothing that meets these requirements has yet been developed, though according to Ruth Eigenmann, Vanni Eigenmann's aunt who is very active in the Foundation, the most promise lies either in a radiometer to measure the thermic radiation of a human body, or in a special form of radar. At the moment, however, radiometers are in their early stages of development, and more knowledge is needed about the propagation of very short electromagnetic waves in different kinds of snow for further progress to be made. The natural temperature variations in an avalanche field are, in any case, considerable as a result of the mixing up of various snow layers, and so the signals received from a human body will always be very weak. The radiometer will therefore have to amplify these signals greatly, and this will certainly influence the weight and size of the instrument. In turn, if the instrument is cumbersome, its difficulty of transport will pose problems and limit its speed of search. And a sophisticated and highly sensitive radiometer, with sufficient range to be effective, will almost certainly be costly.

Radar was one of the first principles tried by the 'Vanni Eigenmann' Foundation, and the initial results in the mid-1960s seemed to hold considerable promise. As most people know, radar consists of emitting electromagnetic waves which, when they strike a target, are reflected back as echoes. These can be displayed to show the position of the target. The human body has a different dielectric constant from that of snow; that is to say that when electromagnetic waves strike a body they are affected differently from when they strike snow. In theory, therefore, a body buried in snow would be a reasonable radar target. In practice, however, the echoes coming from an avalanche field are confused by the ground below the snow, by rocks, by free water, etc. The result is that, so far, the progress with radar has been disappointing.

The future for radar in avalanche rescue, if there is one, may be in a tridimensional system which will show the shape of targets and thus allow their identification. This would call for complex and relatively heavy equipment, and the natural and fastest way to carry such radar would be in a helicopter. But helicopters cannot fly in all weathers, so new problems would be posed. Even fully portable radar would have problems in bad weather because the monitor screen would be difficult to read. Alternative audio signals through earphones would require considerable training before a man, exerting himself to struggle over an avalanche and moving the apparatus about at the same time, could interpret them.

Presumably, the problems connected with radiometers or radar will one day be solved, but the harsh reality of the difficulties that have emerged during the experimental work of the last 10 to 15 years has led to two basic conclusions: firstly, if such a device is ever developed, its use will be only by formal rescue teams because it will be too expensive and sophisticated for widespread distribution; secondly, the idea of having the victim equipped with a device to facilitate finding him is worth pursuing, even if it did not seem to be so 15 years ago.

However, this being said, we have come a long way from the idea, first mooted around 1960, that skiers should all carry special magnets so that magnetometers similar to the Förstersonde used to locate Vanni Eigenmann's body could be brought into general use. The complexities of ensuring that every skier be equipped with a special magnet which he had either purchased, or which had been built into his equipment, would have been overwhelming. In the event, the problem did not arise because tests with magnetometers proved them to be difficult and slow to use. They have virtually dropped out of the scene therefore, even if they can be useful for locating large ferrous masses such as buried vehicles and snow cats.

It was the work of a Swiss called Bächler in the mid-1960s that led to a technical breakthrough in respect of rescue systems requiring that the victim carry a complementary device; and it also led ultimately to a change in attitude towards such systems. Bächler built a small radio transmitter that, when buried in the snow, could be located quite easily by anyone on the surface with a portable, long-wave, radio receiver.

Initially, not many people were impressed by Bächler's work, but gradually the aforementioned change in attitude towards rescue devices crept in. It was reasoned that many avalanche accidents take place so far from the nearest rescue headquarters that even the most highly effective victim detector would probably arrive too late to find the victim still alive. (In effect, this same situation applies also to avalanche dogs.) Therefore, some sort of device that could be carried by every member of a party and which would allow for rapid self-help would indeed be useful. And this would be in addition to radar or radiometers for use by rescue organizations when looking for the casual off-piste skier who has got himself buried but is not carrying any special device to facilitate his rescue.

Many avalanche accidents do indeed involve organized ski-touring groups, members of safety and rescue organizations, military patrols and others who are sufficiently aware of the dangers they are to encounter, and also sufficiently disciplined, to carry special equipment that will ensure prompt rescue of any buried member(s) of the party by their more fortunate, unburied, colleagues.

Thus it is that, nowadays, we tend to think in terms of, on the one hand, self-help rescue systems for use by organized groups, and, on the other hand, of more complex victim detectors for use by rescue teams.

And Bächler's early work was quickly taken up by other individuals and companies with the idea of producing a compact, light-weight transceiver—a radio that can transmit and receive—at relatively low cost for use as a self-help system. The result is that, since the late 1960s and early 1970s, there have been several effective devices on the market. In addition to the Bächler, there is the American Skadi, developed by Dr. Lawton; there is the Swiss Autophon, the Austrian Pieps and the British Skilok. All of these devices weigh from 200 to 400 grammes, vary in size and shape from a large cigarette pack to an oblong rod 19 cm in length, and work on long wave; the Bächler is still a transmitter only and requires a long-wave receiver to complement it, but the others are all transceivers.

The way these devices are used is as follows: the carriers of the transceivers ensure that their sets are in the transmit mode before setting out into danger, and that they are functioning properly. Should an avalanche occur and one or more of the party be buried, the remainder immediately switch their sets to the receive position, twisting them about, and if necessary moving around until the bleep, bleep, bleep from the buried transmitter is picked up. The homing in that follows is relatively simple and does not call for very special training, for the ferrite rod antenna in the receiver acts as a direction finder. Twisting the receiver left and right, the operator will note that the signal is strongest when the receiver is facing in a certain direction. Having identified that direction, he then turns the volume down until it is barely audible and begins to walk a straight line. The strength of the signal should increase as he walks and if it does not, it means he is walking directly *away* from the transmitter. A 180° about-turn and retracing his steps will produce the desired strengthening of signal as he walks. At a certain point, however, the signal will begin to fade again. Returning a few feet, the operator establishes the point of maximum signal, reduces the volume until it is barely audible again, twists the set about until he finds the maximum signal strength position and starts walking towards it again. He repeats this until movement in any direction weakens the signal. He is then standing directly over the buried transmitter.

This process, although long to describe on paper, takes only a few minutes to accomplish in practice. In a series of independent tests, the average search time varied from 7 to 22 minutes according to the make of transceiver. The same tests included some severe treatment of the various devices and, overall, the Autophon was judged the best. (Incidentally, the Swiss Army has bought many thousands of them.)

There can be no doubt that such transceivers are destined to become standard equipment for organized parties of ski-alpinists, safety and rescue teams, etc. They are not cheap, and I hesitate to quote a price because inflation will probably make it obsolete before this is printed. However, as a rough guide, one should reckon with not getting much change out of $100. The Alpine Ski Club of Britain has launched a hire service for Autophons. They bought a number with money made available by a benefactor and, by hiring them to ski-touring groups at a better than break-even rate, they hope to accumulate funds that will allow the purchase of more. This seems to be a good formula, worthy of imitation by other clubs.

The 'Vanni Eigenmann' Foundation considers existing transceivers to be only the very early beginnings of such devices and is promoting research into microwave sets—as opposed to long wave—which would result in even smaller and more efficient units. They are concerned also with making the devices completely foolproof.

There is some justification for this attitude, for the famous adage about aircraft design applies here also: the adage says that if an aircraft and its controls are designed in such a way that they allow even the slenderest possibility for pilot mishandling, one day a pilot *will* mishandle them. At least two cases of misuse of avalanche rescue transceivers have already occurred in the few years they have been available. The first occasion was in 1973 in the Austrian Tyrol, when a skier with a Pieps transceiver participated in the rescue of another skier who was also Pieps-equipped. He then went his way, but forgot to switch his Pieps from the receive mode that he had used during the rescue to the transmit mode he needed to protect himself. And sure enough, he was caught in a second avalanche, and the rescuers could not find him quickly because his Pieps was not transmitting. Then, in January 1975, an Austrian safety patrol equipped with Pieps transceivers was carrying out an operation with explosives for avalanche control when one of the team was caught. His team-mates very quickly found the Pieps transceiver; but that was all they found because it had come out of the pocket of his jacket as he was tumbled over and over in the avalanche.

The 'Vanni Eigenmann' Foundation takes the view that if such gross errors can be made by experts, the devices as they stand need a great deal of modification if they are to be of any use to the ordinary skier. For this reason, the Foundation has designed systems to build transceivers into standard ski equipment, has looked into the question of batteries that might last the full length of life of the transceiver, and is also examining means of turning the transceiver on automatically when its carrier is caught in an avalanche.

Personally, I believe that such attempts to adapt transceivers for the mass of skiers can never succeed. Nor can any device which is relatively costly and which the holiday skier with a bent for deep snow has to make a conscious decision to buy. For a lot of avalanche accidents involve good downhill skiers who enjoy deep snow and go looking for it off the beaten pistes. Many of these skiers know next to nothing about snow and avalanches and would almost certainly not buy a transceiver or any other special device, simply because they never imagine that they might be caught in an avalanche.

On the other hand, serious ski-tourers *are* usually aware of avalanche risks and would therefore buy or hire transceivers to equip a party. And since ski-alpinists usually go into remote areas, it is even more important that they be equipped to rescue each other, rather than having to wait hours for rescue teams to arrive. It seems to me that the better transceivers made today are really quite satisfactory and that the main drive should be to promote their use among ski-tourers, and groups who work in the mountains. Such promotion should also of course stress the need to keep the devices in order, change the batteries at the appropriate intervals and generally use the devices properly. No doubt the existing transceivers *can* still be improved, but to me this is of relatively small importance. On the other hand, were it possible to reduce their price so that more groups could use them, that would be beneficial.

Rescuing the average downhill skier who gets into an avalanche presents a whole series of different problems, if one accepts the point of view I have tried to propound to the effect that it is unlikely that he will buy a device such as a transceiver. In this case, the rescue teams must be equipped with effective radar or some similar device; and there is one very good idea that the 'Vanni Eigenmann' Foundation has put forward in connection with radar. As has been mentioned earlier, there are numerous technical problems that still hamper the development of a radar system for locating a human body in snow, but possibly some of these problems could be overcome if the body carried a metallic radar reflector. During the Second World War, radar systems were quite effectively jammed by bombers releasing hundreds of strips of metal foil (known by the quaint name of 'window') in the sky. The foil provided so many radar echoes that the operators on the ground had difficulty distinguishing the wheat of aircraft from the chaff of 'window' that was cluttering their screens. To build a metallic foil backing into every lift- and cable-car ticket or ski-pass or season ticket would be perfectly feasible. And if indeed such a reflector would simplify the problem of designing an effective radar system, with a range of at least 50 feet, this would seem to be a very promising area to explore.

The 'Vanni Eigenmann' Foundation does not only interest itself in areas of high-flown technology; the more mundane question of digging the snow away from a victim once he has been located has also claimed its attention. Walter Good of the Swiss Snow and Avalanche Research Institute was the first person to raise this problem seriously; he pointed out, at the Solda Symposium organized by the Foundation in 1975, that it takes 40 minutes to move one cubic metre of snow using the heel of a ski, the usual digging device employed by a skier when digging out a friend. Good raised

27. An uprooted tree smashed through a house at Davos in January 1968

28. Airolo after the Vallascia avalanche of February 1951. Photograph taken from the church tower

this issue as a caution to the Foundation which, at the time, was exploring the installation of transceivers in the binding of skis. And the Foundation was quick to take the point: it was of little use locating the transceiver quickly if it were near to a victim's feet and many more minutes would then pass before the rescuer was able, with the heel of a ski, to free his head. The Foundation therefore returned to the idea of affixing the transceiver as near to the head as possible—i.e. on the thorax—and began also to look into the improvement of cheap and light shovels that could be carried in order to cut down the digging time.

Other areas of research which are receiving attention are balloons tied to avalanche cords which would inflate automatically when an avalanche occurred, and balloons attached to the body, and clothing too, that would similarly inflate automatically and keep a victim on or near the surface of an avalanche. I understand the need to explore all possibilities, but I cannot get very enthusiastic about such devices. For they require a conscious decision to purchase them, and they have dubious advantages over transceivers. But perhaps the effectiveness of the traditional avalanche cord would be improved beyond recognition were a hydrogen-filled balloon to be tied to its end. Thus it would float in the sky, pulling the cord up and out of the way; and presumably the balloon would remain tethered in the air above an avalanche field thus making it easy to locate the victim.

The existence and work of the 'Vanni Eigenmann' Foundation should be very much appreciated by all those concerned with avalanche safety and rescue. The symposia it organizes bring together avalanche-rescue professionals with electronic engineers so that a dialogue can take place and each understand better the needs and possibilities of the other. Without this mutual understanding, little constructive work can be done. In addition, the international nature of the Foundation's activities is creating the information flow between countries that will lead to agreed standards and specifications for new equipment as it is designed and proven effective. Even if the Foundation financed no research work itself, these contributions would be invaluable. But the Foundation has also been generous with financial backing for promising research and development activities and has contracted out such work to several organizations. The time and resources poured into this work by Dr. Gino Eigenmann, and by his wife Ruth, in memory of their nephew

Vanni, deserve the highest recognition. They only ask that anyone developing an effective rescue device with their help stamp the initials V.E. on it, and that anyone developing effective devices, with or without their help, readily agree to licensing arrangements in other countries in order to ensure rapid and quantity production and distribution.

10
Avalanches and Explosives

The anti-avalanche constructions which have been mentioned on occasion in this book, and which will be dealt with in full in the last chapter, are incredibly costly. They are so costly that their installation can only be justified to protect permanently inhabited areas. But it can also be that temporary protection against avalanches is required; this is so, for example, when a workers' hostel is put up in connection with a dam or tunnel building project which may only last three to four years. As a means of protection in cases of this sort, it is usual to release avalanches intentionally with explosives, bringing down the snow masses before they can accumulate sufficiently to cause a disaster. During a blizzard it may be necessary to release small avalanches repeatedly if a large and destructive one is not to form.

And for non-essential lines of communication, the use of explosives is particularly suitable as a means for controlling avalanches. The road, railway, ski-run or path can be closed to traffic during times of danger and re-opened once made safe, or proven safe. I say proven safe because the explosives may not necessarily release any avalanches. If this is so it can *usually* be assumed that conditions are not particularly hazardous and the explosives, far from being wasted, have been used as a test for the level of danger. In addition, an explosion in snow has a local hardening and therefore stabilizing effect. This is particuarly so when the explosion is deep in the snow. American experiments have shown that 2 pounds of tetrytol exploded on the surface affect an area only 6 feet across, but that the same charge detonated on the ground under the snow can, dependent on the snow's original density, harden an area several times as great. Even if no avalanche is released, therefore, an island of stable snow is created by the explosion.

Though the intentional release of avalanches was used with dire effect as a means of *taking* life in the First World War, the method was not used to *protect* life and property until 1933 when Ingenieur

Zimmermann, director of the Bernina mountain railway, began tests with rockets and mortars. The idea, however, can be traced much further back, to the Middle Ages in fact. In 1438 a Spanish knight-errant called Pero Tafur, engaged in a series of travels which he later described, made his way via the St. Gotthard Pass to Basle.

To cross the Pass he used a trailer 'like a Castilian threshing machine', a sledge-like contraption, drawn by oxen. The noble knight sat on the sledge and his horse was led behind. He remarked that the advantage of travelling in this way was that if anything fell over a precipice it would be the oxen rather than himself or his horse. Before passing any places that they thought menaced by avalanches, the party discharged firearms to release them. Pero Tafur does not relate whether they brought any down but, in the light of modern knowledge, it is very unlikely that they did. Even today, problems remain as to the most effective, cheapest and safest way of releasing avalanches intentionally. These problems will be examined in more detail later.

Following on from Zimmermann's experiments in 1933, Christian Jost of the Parsenndienst began trials in order to make the ski-runs of the Parsenn safe after fresh snow-falls. Then, with the Second World War on the horizon, the Swiss Army also took a sudden interest: the ability to safeguard lines of communication was to them all important, and the haunting experiences of the First World War in the Tyrol were goading them on to find a solution.

With the aid of the Swiss Army, therefore, a variety of methods for delivering explosives into avalanche slopes were tried: among them were rockets, catapults, grenades, field guns—even explosives carried along a cable, and charges placed in summer to be detonated by an electric circuit in winter. But two means proved superior to all others at the time: the 3-inch mortar (see photograph 22), and the simple but effective method of throwing a tin can filled with explosive into the slope by hand. More recently, the anti-tank bazooka has also been brought into use in Switzerland.

The technique of using hand-thrown charges was developed by the Parsenndienst in the late 1930s and is still in widespread use in the Alps today; for the 3-inch mortar has two major drawbacks: it is difficult to transport and takes a long time to set up. True, it divides into three parts—barrel, stand and baseplate—but each part weighs about 45 pounds and each bomb weighs over 7 pounds.

To lug a mortar and a dozen or so bombs about the mountains at high altitude then is hardly an attractive proposition.

This has meant, in effect, the relegation of the mortar to permanent positions in the valleys, or to buildings in the mountains from where it can be used to protect roads, railways and other installations. In this application, the many advantages of the mortar show up well. It has ample range (nearly 2 miles) and can be fired in fog or a snow-storm provided that blind-firing data has been prepared in advance. This means that a set firing position and a mock target a few yards from it have been chosen for firing at a given slope. The error in degrees between the mock target and the real target has been measured in fine weather, and to fire in bad weather the mortar is aimed at the mock target and pivoted this amount. Charge and elevation have also been calculated in advance, and the latter is set by using the spirit levels on the mortar as datum. Other advantages of the mortar are the parabolic trajectory of its bombs, which makes ricocheting less likely, and the sensitive contact fuse which ensures detonation of a bomb even if it lands in very soft snow. The mortar is also accurate and simple, and the bombs, at about four times the cost of hand charges, are not overly expensive.

But, of course, cheapness and ease of transport are very important, and it is mainly for this reason that hand-thrown charges have become the commonest means, at least in the Alps, for controlling avalanches in ski-areas. The types of charge used will be discussed later, but in the German-speaking parts of Switzerland the tendency is to continue to use dynamite packed into tin cans, the original type of charge developed in the late 1930s. In German, they are called *Sprengbüchsen*, literally translated as 'blast cans'. Their main snag is that the placing of them in the slopes is a skilled and hazardous undertaking.

*   *   *

After every snow-fall of any magnitude during the winter the blasting patrols set out to make the ski-runs safe for the thousands of skiers who will later use them, but many of whom meanwhile champ and fret at the 'Run Closed' signs, ignorant of the menace which is being removed on their behalf. The Parsenndienst has vast experience of this work and the patrolmen carry out many blasting operations each winter. I participated in a fair number while I worked with them. Since then I have organized and carried out many myself in Italy. But one in the Parsenn area is engraved on my memory, that of February 13th, 1964.

It had snowed spasmodically for 48 hours and Christian Jost had closed the ski-runs endangered by avalanches on the morning of February 12th. During that day and the following night the snow fell more thickly and on the morning of the 13th, as we took the first train to the Weissfluhjoch, we know that the blasting patrols would be sent out. It had almost stopped snowing and the visibility was fair.

As we clumped into headquarters and unpacked the rucksacks and rescue sledges used the previous day, the night duty patrolman was already talking to Jost on the telephone, telling him that the north-west wind had been gusting to 80 kilometres per hour (50 m.p.h.) during the night and that 40 centimetres (16 inches) of fresh snow had fallen. Jost told him to have 30 *Sprengbüchsen* made up as a start; 16 inches of new snow with a 50 m.p.h. wind would obviously create a dangerous situation, especially as the underlying strata were composed almost entirely of cup crystals, owing to the thinness of snow cover and the very low temperatures that had prevailed earlier in the winter.

We packed sticks of dynamite broken in half into tin cans provided by the restaurant below, eight sticks into each. We cut fuse into 60-centimetre (2-feet) lengths and crimped a detonator on to the end. Then we slipped the detonator into the yellowish paste of the dynamite. We curled the fuse down into the cans just leaving a protruding end, packed paper on to the dynamite to hold it in place and squeezed the top of the cans closed. During the war the Swiss Army had special tins with holes in the lids from which the fuse stuck out, but ordinary preserve cans serve the purpose.

Jost rang up again from his office in Davos and issued his orders. Beusch and Waber to Parsennfurka and then into the Schwarzhorn east flank, Waber radio-operator. Caviezel, Schwendener, Fraser, via the Small Watershed into the Schwarzhorn east flank from the south, Schwendener radio-operator. Vogt, Bebi and Wehrli as safety patrol to close off the area and observe the blasting patrols, Wehrli radio-operator. Unold and Schwarz to Gipfel North—and so on until just a nucleus of men would remain to come to the rescue of a patrol in trouble. All were to carry avalanche cord, avalanche shovel and sounding-rod.

We gathered up our equipment and packed it into our rucksacks with the cans of dynamite, tied our avalanche cords around our waists and tucked the rolled-up ball into a pocket. Not much was said as we prepared ourselves, for there is always an apprehensive expectancy tempered with suppressed excitement before a blasting operation, however many one has done.

The east slopes of the Schwarzhorn, which we were to blast, rise steeply above the first part of the famous Küblis and Klosters ski-runs. The Schwarzhorn is like a cockscomb, and during a blizzard the wind piles the snow into the corrugations along its flank from where avalanches can plunge down and sweep across the ski-run. These avalanches must therefore be released, or the slopes proved safe, before the run is opened.

To reach the slopes we climbed up an exposed ridge where the wind had grooved and carved the snow in patterns. Meanwhile, our safety patrol skied on down to a safe vantage point from which to observe us. Cloud drifted about, dropping its clammy shroud over us from time to time, and a few fitful snow-flakes loitered down. We wondered whether we would be able to blast, for no patrol is allowed to do so in bad visibility. In the first place, the safety patrol cannot watch them, and secondly, a skier who has disobeyed the 'Run Closed' sign might be in the way. This had almost happened four years earlier when mist momentarily obscured the view. A warning over the radio link from the safety patrol prevented the fuse being lit in the nick of time. The two skiers were lucky, for it was the first year that radio was in use by the Parsenndienst; a year before they might have been buried.

Georg Caviezel, the leader of our patrol, stopped when we were high enough on the ridge to begin our traverse into the slopes that were to be blasted. He threw his avalanche cord out behind him, and Heini Schwendener and I did the same. Caviezel was in his late 40s and one of the older patrolmen at the time. He had an impish sense of fun which led to much horseplay and laughter in leisure moments, and his colleagues were very fond of him—even though they poked fun at his baldness and his predilection for reading westerns. He was a member of his village music club and his yodelling was a feature of any party he attended. For several winters, the Schwarzhorn east flank had been his special blasting domain.

He told Schwendener, who was carrying the radio, to call the safety patrol and inform them that we were about to begin. Then, with about 50 yards between us, avalanche cords trailing, and wrists slipped free from ski-stick straps, we traversed the first gentle slope until we reached a rock outcrop with a cornice extending downwards from it. Georg Caviezel looked at the cornice, took two steps up it, stopped and came down again. He thought for a moment and looked at the rock-studded gully below down which he would be carried if the cornice broke.

'I don't like it,' he said. 'Give me a can.'

I unpacked one from my rucksack and passed it to him. He split the end of the fuse with his thumb-nail to expose the powder, lit it, and tossed the can out of sight just over the cornice (see photograph 24). We huddled back against the rocks, and Heini Schwendener told me to hold on because we might be dusted from above.

The 90 seconds which the fuse takes to burn always seem an eternity, but finally the explosion shattered the stillness and reverberated among the crags. A rustling noise followed which grew in volume until it became a loud and evil hiss before gradually dying away.

'There it is!' said Schwendener, and down in the valley a tongue of moving snow shot into view. It swept across the ski-run and curled away into the distance, slowing as it went.

The radio bleeped as Hans-Martin Wehrli of the safety patrol called us.

'A good one,' he said. 'It's run at least 400 metres, right across the *piste.*'

We climbed over the cornice which had blocked our view of most of the avalanche and which was still in place, but we now judged it reasonably safe. The soft-slab avalanche had broken away from an area about 50 yards wide and had left a fracture line about 3 feet high. We skied across on the foundation of friable cup crystals, the new snow layer having slid away entirely.

Each wind-sheltered depression of the mountain side was pregnant with drifted snow. The patrolmen know their area so well that they can tell at a glance what depth of new snow is lying in a hollow, and, at the edge of each, Georg Caviezel threw a *Sprengbüchse*. Then we retired a few yards to watch the result; the first four explosions were all positive and produced large avalanches. It is very exciting to set hundreds, or thousands of tons of snow in motion and watch it rush away down the slope; and we were delighted by each success.

But the fifth charge exploded and merely left a crater in the snow. We stood in uneasy silence, for this is the situation that a blasting patrol dislikes: a negative result when it is known that the snow cover is unstable. One is in a quandary whether it is really safe to proceed or not. (After a positive result it is almost always safe to cross the avalanche path.) Beyond the crater, some 40 yards away and slightly below the rib we were on, there was a small rock outcrop, and Georg Caviezel told me not to move until he had reached it. Then, with a glance to check that his avalanche cord was trailing free, he set out.

I watched him carefully as he skied away. He had gone about 30 yards when I was disturbed to see that he had run into deeper snow; it welled up around his thighs and he was cleaving a big trench as he went. At that moment a line tore soundlessly along the snow above him, like a zip-fastener opening the slope. In an instant the snow masses along the rocks above burst into movement and pounced like something alive towards Caviezel.

'Look out!' I shouted. He glanced over his shoulder and turned his skis further down the slope, in the hope of skiing into safety beyond the rock outcrop. But as he swung below it, the snow in front of him broke away as well.

I saw him thrown outwards and down the slope and then the swirling snow-dust hid everything. I once thought I saw him come to the surface but when the avalanche stopped and the snow-dust began to settle, there was no sign of him—only an enormous expanse of tumbled snow at least 200 yards wide. The avalanche had struck like a viper and gone hissing down the slope at incredible speed. It had swept clean like a monstrous wave, apparently dissolving Georg Caviezel into nothing. I was awe-struck by the power of what I had seen, and momentarily horrified; but then a strange calm came to me as I realized that a friend's life was at stake, and that only our actions could save him.

I skied quickly to where I had last seen Caviezel—his disappearance point—which I intended to mark with a ski-stick; but then the incongruous thought struck me that it would be a long climb back up for my stick and that, anyway, we were certain to find him through his avalanche cord. Schwendener and I skied slowly down, sweeping from side to side as we search. The safety patrol were climbing desperately towards the avalanche field.

We could find nothing, and after a few minutes I was really afraid. Then suddenly we saw a black object. We rushed to it and found that it was the gloved fingers of a hand, reaching through the snow, clenching and unclenching convulsively.

Beside the arm at the bottom of the hole in the snow was a mouth, spluttering and gasping for breath; and at each exhalation the surrounding snow was flecked with blood. My immediate thought was that Caviezel must have internal injuries. Carefully we scooped the snow away from his head, but it was impossible to do so without occasionally knocking some down, at which he shouted protests. I put my hat across his face so that dislodged snow would not block his mouth and nostrils, but he yelled for me to take it away. He later said that all he wanted was air and light, and that even my hat across his face was terrifying.

The safety patrol arrived, out of breath from their climb, and Georg Caviezel's head and torso were soon free. I stood back to take photograph 25 and suddenly, a thought which had lurked at the back of my mind popped up.

'Where's his avalanche cord?' I asked. The work of freeing Caviezel's legs stopped, and we all looked around. Someone grabbed the cord at Caviezel's waist and pulled the whole length out of the snow. And we realized then that not a single inch of the 30 yards had remained on the surface as a telltale.

The implication was horrifying; had Georg Caviezel been buried 6 inches deeper his fingers would not have reached the surface, and he might well have died before we could find him in such a large avalanche.

He was suffering from shock so we did not dwell on the matter then. He had been bowled over and over and hit on the mouth by a ski-edge, hence the blood-flecked snow. He remembered coming to the surface once and he tried to swim. Then, as the avalanche slowed, he felt the pressure increasing on his body. He was lying on his back when it stopped, with one arm blocked immovably beneath him and the other in front of his mouth and nose to try and win some breathing space. He thought he was being crushed to death, and the pressure on his chest was so great that he could not breathe. With all his strength he was just able to shrug himself enough space to fill his lungs.

He was buried for four minutes but, in the terror he freely admitted to, he said he thought it at least 15. He had lost his hat and ski-sticks and his head ached with cold. Two days later, after resting quietly under a doctor's supervision, he was back at work. His final comment on the incident: 'What a horrible death it must be!' To hear him say it, and to see the look in his usually humorous eyes as he lived those four minutes again, would deter anyone from taking the comment light. Yet Georg Caviezel was only 18 inches deep.

The complete burial of the avalanche cord caused a great deal of concern at the time. It was a couple of years before the aforementioned tests with small sandbags attached to avalanche cords and there was—prior to those tests—much trust placed in the simple red cord. Caviezel's avalanche broke in two parts and there was also a fair amount of snow-dust in the air above it. These factors doubtless caused the cord to be buried. However, when I tackled Christian Jost on the topic of the effectiveness of avalanche cords the next evening, he admitted that during his own seven avalanche accidents over the years, his cord had once been completely covered. In the other six cases, part of the cord had remained on the surface, but so had part of his own body. (His craftiness prevented him ever getting into a full-blooded avalanche accident of the sort that Caviezel suffered.) Jost felt that belief in the avalanche cord was fundamental to the confidence of his men, so he swore his companion to secrecy after the accident in which his cord was totally covered, and only divulged the fact to me when I told him of Caviezel's case.

There is a lesson to be learnt from my behaviour after Caviezel had been swept away: I failed to mark his disappearance point which I had seen more clearly than anyone. Yet, in the absence of other indications of his whereabouts, this information would have been instrumental in locating him. Had Georg Caviezel died in that avalanche because we could not find him in time, I would have been partly responsible. The thought was deeply disturbing and I resolved that if I were ever involved in an avalanche accident again, I would follow the proper procedures to the letter.

After Georg Caviezel had been dug free he was sent home with a colleague to rest. Heini Schwendener and I climbed up again and continued blasting, though with reduced aplomb. And the extreme cautiousness we had suddenly acquired greatly increased our consumption of explosives—we hurled two cans into any snow that looked remotely suspect before crossing it!

Perhaps some ideas has been given as to the danger of blasting avalanches with tins of explosive. Georg Caviezel was not the first, and nor will he be the last, to be carried away and buried while engaged in this work. It has already happened to most of the Parsenndienst patrolmen and to many others elsewhere. Sooner or later, they are all duped by the treachery of the snow, probably by a patch which has withstood the blast of eight sticks of dynamite and is waiting to pounce on the first man to set foot or ski on it.

The Parsenndienst are rightly proud that none of their men have been killed, but patrolmen elsewhere have been. Only the meticulous attention to detail of the Parsenndienst has prevented a fatality, and it is interesting to note, in passing, that the majority of the safety precautions they carry out are equally practicable for ordinary skiers: never ski alone off a marked run; wear avalanche cords when there is a possibility of avalanches or better still carry transceivers as does the Parsenndienst now; have sounding-rods and lightweight collapsible shovels in the party; and, most important of all, know what to do in the event of an avalanche.

But even with the best of safety precautions there remain imponderables like the chance of being killed by a fall over rocks. Paul Sprecher, one of the original Parsenndienst patrolmen, was fortunate once. He set out with Georg Caviezel to blast the same series of slopes on which Caviezel was to have his own accident some six years later. From the shelter of a rock Sprecher threw the first charge, but no avalanche resulted. Very cautiously he set out to cross the slope while his colleague watched; but he had not gone far when a peculiar instinct stopped him. As he said later with a laugh: 'The place smelt avalanchy.' He told Caviezel that he was coming back, turned round, and took two paces.

Suddenly Caviezel appeared to him to be moving uphill. The snow was sliding stealthily away under Paul Sprecher's skis. For the first few yards he remained standing, and then he was flung over a large rock outcrop. The sky went dark and he was being tumbled and rolled in the choking snow masses. 'I don't think this one's going to get you, Paul,' he thought; and with his mouth clamped shut he made strong swimming movements, which succeeded in keeping him near the surface. When the avalanche stopped he was facing uphill in a kneeling position—praying, he said afterwards—and with a few strong movements of head and shoulders he broke surface. He would have been saved even if buried deeply, however, for 18 yards of his avalanche cord were on the top.

He was lucky only to rick his back when thrown over the rocks, but like Caviezel he felt chilled to the marrow. Paul Sprecher's courage has been mentioned in connection with the occasion when he broke his leg while transporting an injured skier, but it is even more eloquent that immediately after this avalanche accident he continued blasting and completed his assignment. When reporting to Jost by telephone later, he was ordered, despite the normal ruling about alcohol in working hours, to down two tots of *Schnaps* on the spot.

The instinct for danger which Sprecher showed, even though it did not save him on that occasion, is strongly developed in all those experienced in avalanche blasting. I saw it demonstrated in Burk- hard Beusch, the head patrolman of the Parsenndienst. Beusch is compact and small with fair hair. He has a gentle smile and a voice so calm and soft that you sometimes have to strain to hear him; but he has the agility and reactions of a panther. I know, because while carrying our skis down a rocky, icy path in the dark one evening, I slipped and fell. Almost before I had hit the ground Beusch was sitting on my legs to prevent me sliding over the drop. He is a fast skier and he has more strength and stamina than most bigger men.

Beusch and I carried out a special blasting assignment a few days after Georg Caviezel's accident. A tourist had attempted to ski into a dangerous gully, despite notices posted everywhere warning skiers to stay on marked runs. But his time to die had not arrived; for although he released two slab avalanches, he escaped being carried away by them. After the second, even his limited powers of dis- cretion dictated a retreat. Christian Jost was frightened that some- one less fortunate would follow the tracks into danger, so he decided that the whole area must be made safe, even though it is not normally blasted. Called Stutzalp, it is an uninviting series of rock-strewn depressions running down a north slope into a valley some 800 feet below. Each depression, about 30–40 yards wide, was filled with wind-driven snow; and the whole snow cover was extremely unstable.

The new snow had settled in a very compact layer on top of the old fragile cup crystals. As we progressed, our weight frequently overloaded these fragile strata and they collapsed with a *wummmph*. This noise like a muffled explosion, and the movement under our skis as the snow cover settled, were heart-stopping.

I was crossing one slope after Beusch when my greater weight produced the terrifying *wummmph*, and cracks shot away from under my skis. I had barely recovered my composure when the radio bleeped and our safety patrol told us that a large avalanche had broken away below us. About 100 yards from me the slope suddenly steepened and went out of sight. The movement of the snow under me had been enough to start the avalanche on that steeper ground. The cup crystals were so overloaded and the top strata so cohesive that a fracture at one point was propagating over quite large distances.

We released another avalanche 40 yards from us in this way, as well as the five which explosives sent rushing down; but it was the release of the last avalanche which made me marvel at Beusch's instinct. Conditions were so dangerous that we had used more explosives than planned, and having run out, we were contemplating the final nasty little gully.

'Well, we can't leave it,' said Beusch. We threw some snowballs without effect, and then Beusch began to move forward.

He approached the gully carefully, stopping frequently to size up the situation. It made me nervous to watch him, but he seemed calm and completely unflurried. He advanced steadily, and then he slowed down and crouched slightly. Suddenly, for no reason I could see, he leapt lithely backwards. In the same instant there was the usual muffled detonation, and the fragmented slab accelerated smoothly away with a gathering hiss, carrying away the track to where Beusch's ski-tips by now were. He does not know why he leapt back when he did; perhaps there was some preliminary movement in the snow, but he returned to me with his slow smile, quite unruffled, and with only the fire of excitement in his blue eyes to betray his feelings.

Burkhard Beusch and I have often talked about that blasting operation since. He says it was the best he has ever been on, and, truth to tell, we both enjoyed ourselves thoroughly. We only discovered later that Jost had spent the afternoon in a state of acute anxiety. He paced up and down, listened to our radio conversations, and even rang up his Reserve Leaders and told them to be ready to rescue us from the avalanche that he was sure would get us. He never again ordered the Stutzalp to be blasted with *Sprengbüchsen*; the risk was too great, he said.

*   *   *

The control of avalanches with explosives has produced occasions when the explosives themselves have caused an accident. At the Grande Dixence dam a few years ago, a patrolman took his avalanche dog on a blasting sortie. He threw a charge and turned away, without noticing that the dog had run to pick it up. The poor animal was blown to pieces of course.

A potentially dangerous but somewhat humorous incident took place while the Parsenndienst were still experimenting with the best and most economic charge of dynamite to put in a can. They found a giant tin and packed no less than 30 sticks into it to see whether the explosion would release several slopes at once. Paul Sprecher heaved the monstrous bomb into the slope while a party of observers watched from below. Instead of lodging in the snow, however, the can, with a little whiff of smoke from its burning fuse, bounced and rolled jauntily down towards the observers. It was making straight for them, and Paul Sprecher laughed for years afterwards when describing the indecent haste with which they made off. No one was hurt, for they covered much ground in a short space of time; but the practice now, if there is any chance of a charge not lodging because the snow is hard, is to tie it to the end of an avalanche cord, whose other end is tied to a ski-stick planted in a safe place. The only drawback is that each explosion shortens one's avalanche cord by a few inches.

Patrolmen serve a long apprenticeship before taking charge of a blasting patrol, and it is almost second nature for them to carry out the various safety precautions. Christian Jost was always furious if he heard of the slightest inattention in this respect. There is a very small *ex gratia* payment to personnel caught in an avalanche—fright money as the men call it—but the tradition begun by Jost is that the

Parsenndienst only pays up if the safety regulations have been scrupulously adhered to.

In 1961 a patrolman called Stephan Hartmann was climbing to a safe position from which to throw a charge, when he heard a sharp crack and an avalanche broke away above him. He clung to a rock for a few seconds but was finally torn free and buried, though not deeply owing to his delaying tactics. This was fortunate because his avalanche cord was still rolled up in his pocket. He, too, pushed a hand through the snow and was rescued. Doubtless he was as frightened as anybody, but all he received as compensation was a stern reprimand from Christian Jost.

Jost used to tell his men, with a smile, that it is forbidden for them to lose consciousness when caught by an avalanche. He had sound reason for saying this: for example had either Caviezel or Hartmann passed out, they would not, of course, have pushed a hand clear of the snow.

In a little leaflet called: 'Notes on Avalanche Danger' which every patrolman must learn, it says: 'Don't be frightened! To be in an avalanche is a long way from dying in it. Your colleagues will rescue you.' This is in fact, the basis of the remarkable *esprit de corps* which I had noticed earlier in the Parsenndienst but only understood after a few blasting operations. There is a harmony among the patrolmen, a deep-seated regard and affection one for another which has been forged by the knowledge that time and again each man places his life in the hands of his colleagues. His chances of surviving an avalanche accident may depend on their quick but well-considered actions. Although they may poke good-natured fun at each other, they are very tolerant of each other's faults; for each has seen the true value of his colleagues demonstrated in times of danger. Any man who joins the Parsenndienst and who measures up to their standards of willingness, dependability and courage is accepted with warmth into the group. And the esteem and affection they show soon kindle in him that same pride and *esprit de corps*.

* * *

Avalanche control with explosives is widely practised throughout the world today, though in North America artillery is used more

29. A device for measuring the impact pressure of avalanches and their speed

30. The church at Oberwald, Valais, protected by a splitting wedge

31. The church at Davos-Frauenkirch with pointed end designed to split avalanches

than hand-thrown charges. This is doubtless because there are greater quantities of surplus arms and ammunition available than in Europe, and cost is probably less of a consideration. Permanent gun emplacements or towers are generally built, and on these either a 75-mm or a 105-mm recoilless rifle is mounted. Occasionally, the larger of these rifles, which weighs 700 pounds, is mounted on the back of a truck, and it is then the ideal weapon for avalanche control along highways. The 70-mm howitzer is another weapon often used; it was first proved successful many years ago for safeguarding the highways in the Seven Sisters area of Colorado.

Such artillery can be very effective, as one would expect. On one glorious, but admittedly exceptional day, January 24th, 1964, the snow rangers of the Wasatch Forest Service fired more than 90 rounds from howitzers and recoilless rifles at Alta, Utah. They brought down just short of 30 avalanches, and it is interesting to note that one avalanche for three detonations is also the rough average for the Parsenndienst; but it must not be forgotten that safety measures can be achieved without actually releasing any avalanches.

One of the avalanches brought down at Alta that day sent the gun crew running for their lives and buried the howitzer. Another went into the car-park of a ski lodge and buried several vehicles, including a snowcat which the driver, thinking that the firing was finished, had just climbed into. He was rescued unharmed. Two guests at the lodge were outside, also having heard the rumour that firing was over. One became so excited at the approach of the avalanche that he ran full tilt into the swimming-pool. Another of the avalanches released by the rangers destroyed a ski-lift and a 1,000-gallon storage tank. All in all it was quite a day. It followed a storm that, in 66 hours, deposited 50 inches of snow on to an unstable base of cup crystals. Alta, in any case, has frequent and tremendous avalanche cycles.

The lightest of these weapons used in the U.S. is the 75-mm recoilless rifle; but that weighs 167 pounds and so it is even less portable than the 3-inch mortar. It follows, therefore, that if artillery of this nature is to be a practical proposition for avalanche control, there must be sufficient resources available to place a number of fixed weapons in an area—unless there just happens to be a single vantage point from which all the dangerous slopes can be shot at.

Three basic problems emerge in any serious consideration of controlling avalanches with explosives. The first two are technical and concern, firstly, the *means of delivering* the charge on to the slope and, secondly, the *characteristics of the charge* itself. The third problem relates to the *legal and juridical aspects of using the technique* in countries where there are stringent controls on the use of all explosives and where their use for avalanche control has not been foreseen by existing legislation. Let us take these problems singly and examine them more closely.

The question of a delivery mechanism for putting a charge in the desired place on a slope, cheaply, safely and accurately has not been fully resolved. The North American solution of artillery pieces on fixed towers is not generally applicable elsewhere, effective as it is. The Swiss use of portable weapons such as the mortar and bazooka is also dependent upon some peculiarly Swiss conditions. One of these is that in Switzerland all fit men up to late middle age are obliged to spend about two weeks each year in military service, refreshing their knowledge of the military skills they learned during the longer conscription period of their youth, or learning new skills. Thus, it is quite common to find, in a mountain rescue and safety group, men who are fully trained mortar or bazooka crews and who get refresher training from the army every year. When such a crew is available, the army leases mortars and bazookas to the safety organization at very low cost. But the organization has to pay full price for its mortar bombs and rockets, and these—especially the rockets—are expensive.

The ideal device for delivering a charge into a slope must fulfil a variety of requirements: it should have a range of at least one mile, be accurate to within 10 feet or so and deliver a charge equivalent to at *least* two pounds of TNT. It must also be reliable at low temperature and varying humidity, not be seriously affected by high winds, have an aiming system that allows blind firing in poor visibility, and, above all, it must be simple to use, easy to carry and cheap.

The arms industries of the world could design and produce such a device within a matter of weeks, but they do not do so because there is not a big enough potential market. So we have to look among existing military weapons, and the best of these is probably the anti-tank bazooka, for countries where such weapons can be made available for avalanche control.

The bazooka is so light and quick to use that two patrolmen can easily take it to a vantage point and release a series of rockets into dangerous slopes. It is rather short on accurate range, however—about 1,200 yards—and it has no blind-firing aiming system. This could probably be overcome by setting up aligned aiming stakes with hoops on top, or some such device; but the main drawback is that anti-tank rockets have very firm contact fuses and if a rocket goes obliquely into very deep snow it may not always explode.

The problems with military weapons and the dangers of throwing charges by hand—unless a cable car conveniently crosses a target slope and a charge can be lobbed from it—have caused people to look for other solutions. Throwing charges from light aircraft proved too inaccurate, though helicopters are much better. One company in Switzerland makes a special avalanche bomb designed for dropping from a helicopter, but the drawbacks are of course that helicopters are expensive to operate and need good weather.

In France, the old Swiss idea of placing charges in the summer and detonating them as necessary in winter has been revived. A company produces networks of electric circuitry with charges which are buried in the ground and can be commanded from a single control panel. Because of the need to bury the components of the system, it is unsuited to rocky slopes. In addition, the idea of leaving charges in the ground for months on end meets with criticism in many quarters and is actually illegal in some countries. The purists also point out that a lightning strike could detonate the charges: and what about landslides or thieves carrying the explosives away, they ask.

A much more practical solution is, without doubt, the one adopted in Bavaria, West Germany, at Garmisch and Mittenwald. There, small cableways have been built and charges are pulled into position along them by hand or by a small motor. The cables, some of which run almost 2 km, are several metres above the slope. The charge, or several charges at once, are tied to them so that they dangle just above the snow but are far enough below the cable for the blast not to damage it. The German cableway systems are quite elaborate, but simpler devices, like glorified laundry lines running on pulleys, can be built almost anywhere. I have proposed these during several avalanche consultancies, for the advantages of this system are obvious: the patrolmen can stay in a safe place; it is cheap and rapid;

by marking the cable when the charge is in the desired place, it is possible to establish a method for blasting with zero visibility; and finally, a whole string of charges can be strung across a slope and exploded simultaneously by using a detonating fuse between them. Its only minor disadvantage is that the cableways could be considered damaging to the beauty of the mountains, but there are so many cables of one sort or another criss-crossing a modern ski-area that a few more hardly seem important.

Finally, a few remarks on hand-throwing of charges. Of course, by its very nature, it is bound to be a hazardous operation for the patrolmen carrying it out. However, as a result of many blasting operations I have participated in or been responsible for, I am convinced that the operation could often be made safer, and at the same time more effective, than it usually is. Most hand-throwing of charges takes place from ridges, and as long as the patrol can stay on top of the ridge, the risk is very limited. However, it often happens that from the crest of the ridge itself, a patrolman cannot see enough of the slope he is to blast to determine the exact point at which the charge would be best placed; or if he can see it, he is too far away to be able to throw the charge exactly where it should be. The patrolman then has two choices before him: either he throws the charge as best he can and hopes for successful outcome, or he tries to reach a better vantage point, and in doing so often exposes himself to danger.

Many such critical situations could be overcome by installing ring bolts in a rock, or poles in the ground, to which the patrolman could belay himself as he seeks the best position from which to throw the charge. I know from personal experience that being belayed to a firm object by a good climbing rope gives one a great sense of safety, especially as one moves downwards towards the brow of a convex slope, charge in hand, uncertain of where the fracture line might occur. The rope gives one the peace of mind to work properly, looking for the point of maximum stress on the slope and ensuring that the charge lands on it. Yet few organizations have thought to render the operation safer and more effective in this way.

The second technical problem mentioned in respect of avalanche control with explosives was the characteristics of the charge itself. It should be stated at once that most of the knowledge in this field is empirical and heads of safety organizations have widely varying experiences and opinions. Some like to bury charges in the snow when possible; others leave them on the surface; others explode them in the air over the snow, if they can. Some believe in large charges, some in small charges, and many different types of explosive are used. Without doubt some scientific research on the subject would be very helpful and, fortunately, the Swiss Federal Institute for Snow and Avalanche Research has quite recently begun to study the situation.

Some of the basic facts that are known are as follows: the explosive used should have a rather rapid detonation rate in order to set up a strong shock as simultaneously as possible over a wide area of slope. Normally, explosives with a detonation rate between 3,000 and 6,000 metres per second are considered best. In this category fall explosives based on nitroglycerine (such as dynamite), TNT and seismograph powder. Dynamite is not considered quite as good as TNT, or tetrytol, because it is nearer the lower end of the ideal detonation rate scale. (Dynamite is usually the easiest type of explosive to obtain, however.)

Evidence seems to be building up to show that a charge exploded in the air, about 2 metres above the slope, is the most effective; and whereas a few years ago charges seldom exceeded the equivalent of 1–2 kg of TNT, organizations are now tending to use 3–5 kg as a minimum.

Most organizations still prepare their own charges, but tin-can containers are giving way to polythene wrapping and adhesive tape, or cylindrical cardboard containers of the sort used for drawings, pictures, etc. In France, purpose-made charges in plastic containers can be bought. One particular advantage is that fins can be fitted to the container before it is thrown; the purpose of the fins is to dig into the snow and hold the charge in place, thus overcoming the time spent tying the charge to a cord, and tying the cord to a planted ski-stick, to ensure the charge will not roll or slide away on hard slab.

The research results from the Institute are eagerly awaited. It may well be that different charges and different techniques are required for different snow conditions, but it will certainly be useful to have some sound data available rather than continuing blindly with methods that are often preconceived and influenced by tradition.

The third problem, referred to earlier, is in connection with the juridical aspects of using explosives to control avalanches. In many countries, people who are to handle explosives require special training and a licence. When this is the case, safety and rescue organizations are faced with the need to send staff for training, and this is a deterrent. Far more serious, however, are problems connected with responsibility for possible damage caused by the explosives or by an avalanche released by them. Provided all goes well, there should be no such damage, but unfortunate incidents have occurred. A few years ago, for example, two patrolmen from the Parsenndienst were sent to Wolfgang, near Davos, during a snow-storm, in order to fire some mortar bombs into the so-called Drusatcha, a slope from which an avalanche in 1917 destroyed a train and killed 11 people. They sent four bombs winging through the snow-storm, packed up the mortar and returned by train to Davos.

They were met on the platform by a policeman who told them they were under arrest. They thought it was a joke, but in fact the last bomb had had a faulty propellant charge and it fell very short—eight yards short in fact of someone's garage door. The new house had been rocked by the explosion, the garage doors blown off their hinges and two cars parked nearby peppered with shrapnel. The owners of the house had only just gone indoors after shovelling snow from the drive where the bomb landed. Not unnaturally, there was quite a fuss before the matter was amicably settled.

Even more serious was an episode that took place during the disastrous winter of 1951 in the Alps when more than 700 people were buried by avalanches. A village in the Engadin called Zuoz had decided after the Second World War to use a mortar systematically to control avalanches in two large gullies that came down on either side of the village. During the centuries, avalanches had overflowed these gullies on several occasions and invaded the village outskirts, and so, in 1946, a mortar was procured and a village Avalanche Committee established to decide if and when to fire it. All went well, and small harmless avalanches were regularly brought down, until the great snow-storm of January 1951. Then, a few kilometres away from Zuoz a road worker was swept away, and a second avalanche down the same track swept away some of the rescuers who were looking for him. Finally a third avalanche swept away one of the rescuers of the rescuers. Most of Zuoz's male population, including the Avalanche Committee, was involved in those rescue operations; and the only man left in Zuoz who was directly connected with the use of the mortar was the head of the mortar crew. He had no authority, however, to fire without orders from the Avalanche Committee, which could not be contacted because the telephone lines were down. The head of the mortar crew and the mayor refused to assume the responsibility of firing the weapon. They and the village council debated for hours; some said that it might already be too late to shoot because the avalanches would be enormous, and if nothing was done they might not come down at all. Others were urging immediate action.

Finally a decision to shoot was taken, about 12 hours too late, and the bomb brought down avalanches in both gullies and on the slope directly above the village. This slope had never been known to avalanche since records began in 1598. Five people were killed and considerable damage was done. In addition, some Swiss insurance companies began to write clauses in their policies excluding damage caused by avalanches released by explosives.

Nothing quite so serious as the Zuoz episode has happened since, but nevertheless avalanche control with explosives is still viewed sceptically by some local government authorities in countries where it could and should be used more extensively. In Italy, for example, some local authorities will only allow the armed forces to operate with explosives. In many cases, these forces have insufficient knowledge of avalanches for them to be effective, or they have to be called in from so far away that the critical period has passed by the time they actually shoot or throw a charge. Other authorities in Italy turn a blind eye and let ski-lift companies operate with explosives as they see fit. In fact, the whole situation in Italy, and in some other countries, is very unsatisfactory because of the lack of realistic legislation. Avalanche control with explosives is a very valid technique; in the case of Zuoz, it was the failure of the organization that caused the damage, not the technique. And the same is true for the other cases where it has not given the desired result, for it is a technique that calls for knowledge, skill and sometimes courage from those who employ it.

It is possible, however, that the future of controlling avalanches by starting them deliberately may not lie in explosives at all. As this book was being prepared for press, the Austrian Institute of Large Scale Technology at Gleisdorf announced a completely new approach that it is testing in parts of the Tyrol and Carinthia. Large metal panels are set in the ground on dangerous slopes and can be electrically activated so that they vibrate. By switching them on at regular intervals during a snowfall, small and harmless avalanches can be triggered.

No cost details are available as yet, but my guess is that the system will only be justifiable to protect major roads and railways. Nevertheless, the idea is interesting; time alone will show how practical it is.

II
Protection Against Avalanches

The role of woodland. No doubt the very earliest settlers in the Alps would have liked to protect themselves against avalanches, but understandably they bowed to the idea that these monstrous phenomena were beyond the control of man. It is, on the other hand, less understandable that they should have played straight into the hand of avalanches by ravaging the forests close to their houses, felling those very trees that were affording them protection. It cannot be said that the people were ignorant of the value of timber against avalanches or of the general harmfulness of indiscriminate plundering of forests. The importance of timber preservation was realized as long ago as A.D. 500 when a law was passed in Burgundy to control the felling of woodland; and the first forestry inspectors were appointed in Europe by the French king Childebert III between A.D. 743 and 750.

Documents inform us that by the 14th century the forestry inspectors were much concerned by the way the felling of timber was increasing avalanche danger. A text of 1323 states that if the cutting of the wood behind Bourg d'Oisons (Dauphiné) continued the town would be menaced by avalanches. And Jean Jommaren, a forester of the Oisons area, stated in a document of March 31st, 1346, that 'several people from Oisons and elsewhere are ravaging the woods daily and destroying them to the point where those using the public paths and those residing in the habitations of the area are in danger of perishing at the hands of avalanches and floods'.

In Switzerland the problem was even greater and in the 14th century local authorities began to issue edicts proclaiming certain strategic areas of forest to be *Bannwälder*—banned woods. These orders, having named the wood in question and described its position, usually went on to declare something like the following: 'No one shall remove from the wood anything growing or dead, green or withered; anything lying or standing, small or big. Neither shall anyone dare to remove bark, tinder, berries or cones.' The order ended with the penalties for contravention and the fines were usually high.

One of the first edicts of this nature was that of 1342 in respect of the 'Holz in den Flüen' in the canton of Schwyz. The triangular wood above Andermatt was protected by a banning order in 1397. Without the interdict there is little doubt that this wood would have gone the same way as the other timber in the Urserental; and from the experience of January 1951 it seems fair to deduce that Andermatt could not have survived long without the five-acre patch of forest to protect it. In 1896, when the Swiss Federal Department for the Interior called in all the *Bannwald* orders, there were 322 of them.

The creation of *Bannwälder*, though important and worthwhile, also had drawbacks. Some of the orders failed to prohibit the presence of farm livestock in the wood, and the grazing of young shoots over the years proved disastrous. Even at best, *Bannwälder* were a somewhat negative measure, and the total ban on felling allowed many of the woods to become weakened by old trees, when judicious work by foresters could have kept them strong and vigorous. It is arguable, too, that the role of timber in avalanche defence is not one of straightforward benefaction. Photographs 2a and b amply illustrate how a large airborne-powder avalanche, that gathers momentum on open slopes above the tree-line, can rip down acres of mature woodland. The effect on a village of the normal destructive power of an airborne-powder avalanche, supplemented by flying tree-trunks to act as battering-rams, hardly bears contemplation.

It is unfortunately all too frequent for avalanches to cut a breach through woodland and reach the valley floor. Peaks in the Alps commonly range from 9,000 to 13,000 feet in altitude, but trees, even in the most favourable locations, do not flourish above 8,000 feet. There is therefore plenty of room for avalanches to get under way. A glance along almost any valley in the Alps will show countless vertical gashes in the woods where avalanches come down. Once a large avalanche has cut such a path smaller and more frequent ones will prevent trees re-establishing themselves. The role of timber is best summed up by saying that it gives good protection against avalanches when it is growing densely in an avalanche break-away zone; but it needs protection itself if the break-away zone is above the tree-line.

Early defences, splitting-wedges and diverting walls. The first positive step against avalanches was the placing of large heaps of earth and rock against that wall of the house facing the avalanche slope. In their rudimentary form these heaps were intended to stop an avalanche, but later developments were a little more subtle. Where the house was built on a slope, the space behind it would be filled in completely and levelled with the roof in the hope than an avalanche would pass over the building without doing any damage. Alternatively, the mound would be shaped like a wedge so that the avalanche would be split and pass each side of the building.

Later, it became common practice to build wedge-shaped structures of masonry on the slope-side of the building and these are known as *Spaltkeile*—splitting-wedges. The church at Oberwald in the Valais has one of stone walls 25 feet high and 5 feet thick (see photo. 30), sometimes they were built even higher in an attempt to guard against airborne-powder avalanches. There is a church at Villa, Tessin, where the wedge is the full height of the steeple.

A slight variation of the splitting-wedge was to give the building itself a pointed end like the bow of a boat. The church at Davos-Frauenkirch is a perfect example of this (see photograph 31). Having been previously destroyed by avalanches the church was finally rebuilt in this shape, and with good results. In 1817 an avalanche buried the church so deeply that, until June of the following summer, the only way for the pastor and his flock to enter the building was via the steeple; but the church was undamaged. This incident might lead one to believe that splitting-wedges were a very effective defence, but it is probable that the avalanche which hit the church at Davos-Frauenkirch was running out of destructive power. In truth, the *early* splitting-wedges and mounds were not very reliable.

Perhaps the most striking example of a splitting-wedge is the enormous one that protects the whole village of Pequerel, above Fenestrelle in the Val di Chisone west of Turin. The original part was built in 1716 but it has been extended on several occasions, most recently in the late 1950s. It is now 6 feet thick, about 16 feet high and each branch of the V is nearly 100 yards long. Despite this, a few houses in the village outskirts were destroyed after a blizzard in 1930. Fortunately, the inhabitants had been aware of the danger and had moved to houses nearer the village centre.

Among other early defences were deflection walls set obliquely in an avalanche path to divert the snow away from buildings. The first of such deflection walls may well have been the one built at

Leukerbad following the 1518 disaster. It is still there today, and only a few years ago someone wrote that it was still giving good service. However, one glance at it filled me with relief that Leukerbad has supplemented it with more modern and effective measures; for it looks about to crumble, without help from an avalanche.

The principles involved in early avalanche defences were sound, even if the structures were often inadequate owing to lack of strength. Splitting-wedges are still built today for certain applications, particularly for protecting electricity and cable-car pylons; but they are now massive structures of reinforced concrete. Deflection walls are also constructed, but only at very fine angles to the avalanche path and even then they are heavily built (see photograph 33). And the idea of causing an avalanche to pass over a building is perpetuated by giving the building a strong, lean-to roof and extending it rearward to the slope. Ideally, the avalanche will slide across the roof and do nothing worse than pull off the chimneys.

The disadvantage of all these methods for protecting a house is that, even when correctly executed, they merely turn it into something like a medieval fortress: there may be safety within the four walls, but as soon as the inhabitants venture forth they are at the mercy of the enemy. And nowadays it is hardly practical to be besieged in a house when children have to go to school, there is shopping be done and tradesmen must deliver.

Avalanche control in the break-away zone. In reality, all the methods so far covered have been passive in nature, allowing the enemy to reach the door before offering any resistance. The better solution seems to be to attack the enemy in his lair—that is to say to erect structures in the avalanche break-away zone which will support the snow cover and prevent an avalanche ever starting.

The first attempts of this kind were made surprisingly late, and methods have only reached any degree of perfection in quite recent years. The earliest system was to dig trenches and form terraces in the break-away zone and, according to Johann Coaz, the first of these schemes dates from the 18th century. He based this statement on a document of 1756 which refers to the village of Geschinen in the upper Rhone Valley: after the Birch avalanche had done much damage, the bishop gave special dispensation for the villagers to work on Sundays in the renewal of the ditches and terraces in the avalanche break-away area. Coaz assumed that these defences would then have been about 50 years old if they needed renewing. Terracing of this type was only of limited effect of course. The ditches were usually about 3 feet deep and about the same width, and a winter of abundant snow would smooth them over. But a start had been made in the right direction.

In the 1860s, Johann Coaz himself took the initiative by organizing the building of stone walls in some avalanche break-away areas. These proved successful in general, but suffered from the disadvantage that snow drifted in behind them, as behind any solid object, and in very exposed positions cornices formed on them. Coaz's first walls were free-standing and a later development, suggested by a man called Fankhauser, was to fill in the space behind the wall and make a flat terrace. These were also quite successful but could be submerged in a year of exceptionally heavy snow-fall.

One of the main purposes in building walls to prevent avalanches starting on a slope was to create conditions in which trees could be planted with a good chance of establishing themselves. Then, once the replanted forest was in existence, the walls among the trees could be allowed to fall into disrepair; but of course, any above the tree-line would need to be maintained in a serviceable state (see photographs 34a and b). Reafforestation has remained the ultimate objective of all avalanche control in the break-away zone; and this explains why, in almost every country of the world where avalanche defences are necessary, they come under the jurisdiction of the forestry authorities.

In some parts of Europe, and especially in Austria, the authorities are at present very concerned by the general state of the woodland and the effect on avalanche activity. Dipl. Ing. Alfred Geschwendtner of the Austrian Ministry of Agriculture and Forestry, who is mainly responsible for avalanche defences, reafforestation, etc., says that each year about 2,700 avalanches come down, more or less regularly into inhabited areas of Austria. Of these, about two thirds start below the possible level of the timber-line, but in places where there is at present no timber. With reafforestation these avalanches could therefore be controlled, but this presupposes the installation of snow-support structures to prevent avalanches while the trees are establishing themselves.

Geschwendtner cites what has happened in the Tyrol as an example of a grave degeneration in Austrian forestry and agriculture. A vegetation map of the Tyrol dated 1774 shows that the area of woodland has decreased by half since then, with the highest rate of reduction between 1774 and 1889; and the level of the timber-line has dropped as much as 1,300 feet in some places.

The denuded areas were only of agricultural value for 50 years or so, and then they tumbled down to scrub or weed grasses. The live-stock population was reduced by half from 1850 to 1950, and in the last hundred years the level of production from agriculture and forestry has dropped by a third. The only things to gain in this somewhat dismal picture have been avalanches and torrents: Geschwendtner says that *the area of their paths has increased four times*.

In the Tyrol there are about 125,000 acres that could, and should, be reafforested with the aid of snow-support structures, and the total for Austria is about 385,000 acres. As Geschwendtner points out, each time there are severe avalanche winters the area increases. He believes that a much intensified effort in avalanche control in the break-away zone and in reafforestation is now vital.

It is interesting, when tracing the early history of snow support structures in the avalanche break-away zone, to read of the opposition that there was to this type of defence. Many eminent men argued violently that to banish avalanches would do more harm that that done by avalanches themselves. Without avalanches the permanent snow-line would come lower year by year, they declared, until ultimately the Alps would undergo a second Ice Age. Coaz disagreed, saying that the position of the snow-line was determined purely by climatic influences. He remarked on the very strong melting power of the Föhn wind and illustrated how well this fact was known by quoting two Alpine adages: 'Tonight the wolf (Föhn) will eat the snow' and 'The good Lord and the good sun can do nothing without the Föhn.'

Other people claimed that avalanches, especially the large wet ones in spring, were beneficial because they brought top soil off the slopes and increased the fertility of the valleys by depositing it there. They also said that the large heaps of snow left in the valleys by ava-lanches were invaluable reservoirs of water, delivering their supplies over several months of the summer as they gradually thawed. (This is in fact true and much research is presently going on in Russia and Eastern Europe into the hydrological role of avalanches.)

But, fortunately, no one paid much attention to the cranks, and in October 1902, the Swiss Federal Government passed some far-sighted legislation which assured the future of anti-avalanche constructions. The legislation authorized cantonal and federal subsidies for up to 80% of the cost of the constructions. It also authorized the compulsory purchase of private land necessary for the erection of the walls or barriers, with half the purchase price to be paid by subsidies. In addition, the owner of the land was to receive a sum equal to the total profit he had made from it over a minimum period of three years and a maximum of five, according to circumstances. (The annual profit was taken as the average of the previous 10 years.)

Coaz reported that between 1868 and 1909 there was a total expenditure of 2,048,610 Swiss Francs on avalanche control in the break-away zone. Most of the constructions still consisted of stone walls, although a few experimental steel barriers had also been erected. Coaz remarked that in France, by 1909, protection work against avalanches was also progressing, though in Italy nothing had been done.

But a few years later it was the Austrians who set a trend in avalanche control in the break-away zone: when extending the stone-wall defences above the Arlberg railway, they built heavy wooden fences supported by steel uprights. Fences of one sort or another were soon found to be cheaper and more effective than walls or terraces and by 1939 they had entirely superseded the earlier methods. Fences can be thought of as racks to support the snow cover, though they are also expected to halt any small slides that may happen to start between them.

One could suppose that the design of support structures and their erection in the avalanche break-away zone would be relatively simple matters, but they have proved astonishingly complex. Up to the late 1930s all progress was based on empirical knowledge. There was no basis for calculating in advance the amount of stress the snow cover would impose on a fence; no one knew what were the most suitable materials to use, or how the structures should be deployed on the slope to best advantage. So the men responsible just used their own judgement and any materials conveniently available— anything from railway sleepers and lengths of rail to steel girders and specially cut timber.

But in 1938 Robert Haefeli and his pioneering team, at what was to become the Avalanche Research Institute, produced a basic formula for determining the snow pressure on a fence, and this opened the way for the first calculated designs. Nowadays, so much further knowledge has been and still is being accumulated that a complete guide to the design and installation of snow support structures is issued by the Swiss Federal Forestry Inspectorate. The handbook, compiled by the staff of the Avalanche Research Institute, runs to no less than 60 pages of diagrams, formulae, calculations and text.

Everything is covered: there are formulae for determining the correct siting and spacing of the fences; specifications for materials and designs; instructions for calculating snow pressure on a given slope and at given altitudes; minimum permissible strength margins for different structures over and above the calculated stresses to be imposed on them; details of how to anchor the fences in the slope according to different geological circumstances; how to test the soil's ability to withstand the pressure imposed by the foundations, etc., etc. Indeed, so comprehensive is the handbook that many countries have recognized its value and had it translated. In Switzerland the provisions of the handbook are mandatory where subsidies are to be granted because, as we shall see, this type of avalanche defence is so expensive that no mistakes in planning or execution can be tolerated.

Modern fence-type support structures are of two basic kinds—snow-bridges and snow-rakes (see Fig. 12). Of the two, the snow-rake is slightly superior because all the uprights can be embedded in the ground for extra strength. It also holds the snow better because it has more members running contrary to the stratification of the snow cover. Nevertheless, economics have so far decreed the more common use of snow-bridges, for they require the preparation of fewer foundations.

One of the main prerequisites of a snow-support structure of either type is that it should not influence the way a snow-fall is deposited—that is to say that it should not cause drifting. To ensure this the bars of the structure must be well-spaced, but not so widely that any quantity of the snow cover can slip between them. By a process of elimination it has been found that the compromise gap is 12–15 inches.

The early fences were always erected vertically, but this placed

32. The enormous splitting wedge that protects the village of Pequerel, above the Val di Chisone west of Turin

33. Modern deflection wall at Oberwald, Valais. Note the very fine angle to the avalanche path at which it is set.

34a. Stone wall snow-support structures above the track of the Rhätische Bahn near Albula, Graubünden. Photograph taken in 1907

34b. The same slope as 34a after an interval of 50 years. The replanted timber is well established and the support structures among the trees can be allowed to fall into disrepair

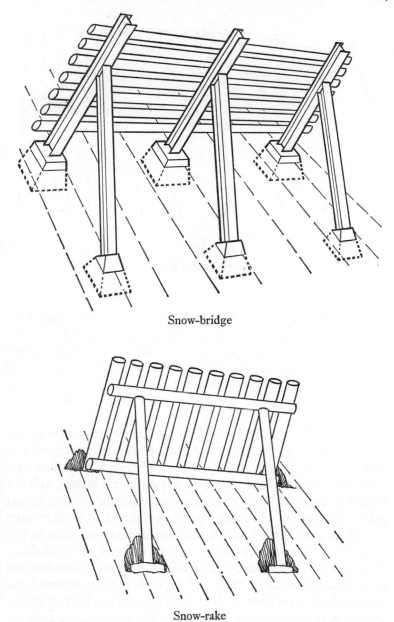

Snow-bridge

Snow-rake

Fig. 12. The basic types of structure for supporting the snow cover in avalanche break-away zones

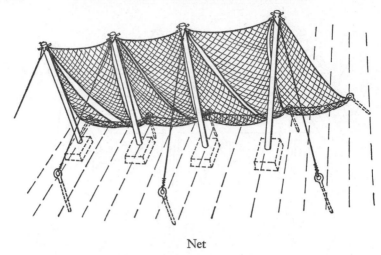

Net

Fig. 12 (*contd.*)

unfavourable stresses on the anchorage and also reduced their effective height (see Fig. 13). It was later found that it is much easier to provide a firm anchorage when the fence is given a slight downhill bias, and it is usual today to set them to make an angle of 100–105° with the slope on the uphill side. As is shown in Fig. 13, this loads the foundations more favourably and makes full use of the structure's height.

A very recent practice is to place the support structures in unbroken lines across the slope; but if an interrupted arrangement is used—as it has been in the majority of schemes in existence—the maximum permissible gap between structures is 2 metres, unless a feature of terrain definitely precludes an avalanche starting at that point. Attempts to cut costs by spacing the units further apart proved false economy because small avalanches were able to run down between the fences tearing out the odd one here and there. Once a breach had been made in this way, larger avalanches tore more units out and so on until, had the units not been replaced, the whole defence scheme would have been reduced to a litter of broken structures at the bottom of the slope. The maintenance costs on schemes that had been skimped in the first place were therefore very high. The tendency nowadays is to spend more initially so as to

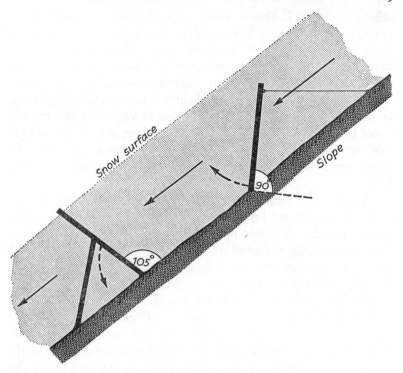

Fig. 13. Early snow-support structure (right) set vertical with resultant loss of effective height, and pressure of snow tending to push base of fence away from the slope. Modern snow-support structure set at an angle of 105° to the slope makes better use of structure height and loads foundations favourably

eliminate the risk to the fences caused by small avalanches starting among them.

The spacing of the fences up and down the slope is equally important. The supporting effect of a structure is only felt upslope for a distance of two to three times the depth of the snow. Using this knowledge, the slope angle, the degree of roughness of the ground and several other factors, there exists a formula for calculating the minimum spacing between the structures up and down a given slope. To exceed this spacing is dangerous because a small avalanche could start, pile up against a fence and so form a ramp for other avalanches. In theory, a succession of avalanches could form a ramp against fence after fence, rendering them ineffective and creating an avalanche slidepath. Of course, structures that are so widely spaced as to support the snow cover only locally will themselves create stresses by giving rise to differential creeping of the snow. These stresses will increase the likelihood of avalanches beginning among the fences. There were a number of very expensive experiences before it was finally concluded that there is no substitute for close spacing of the structures, both up and down and across the slope; hence the recent innovation of placing structures in unbroken lines.

The materials commonly used today for support structures are steel, aluminium and pre-stressed concrete. Wood has a relatively short life and is therefore not used much, except below the tree-line and where the replanted timber will be well enough established in 30 years to take over the structures' task of avalanche control. Experience has shown that when steel is used it has to be of specification that guarantees that it will not become brittle at low temperatures. Indeed, the wide variations of temperature, from the midday sunshine of spring to many degrees below freezing at night, have caused many complications in avalanche defences. Particularly with metal structures, the heat absorption by the bars thaws the snow against them until only the bottom bar is left bearing the weight. And with composite metal and wood structures, the uneven thaw-effect produces unfair loads on the wooden members.

In spring, support structures are often subjected to localized loading. The snow cover is usually less thick than in winter, though its density is higher, and it therefore concentrates its weight on the lower part of the structure. Failure to allow for this caused some expensive mistakes until, in the Swiss Forestry Inspectorate handbook of 1961, a ruling was introduced to cater for such loads. In brief, structures are now built to withstand the total snow pressure spread over them plus a further 25% acting on the bottom quarter of the structure's height.

This brings us to the point that it is often difficult to decide how high a support structure should be. To be fully effective it must, of course, always protrude from the snow cover. From snow-fall records of the area, and making allowance for the extraordinary winter, it is possible to reach a theoretically satisfactory answer. But, owing to wind and terrain effects, snow is hardly ever deposited evenly on a slope, and careful planning is necessary if some structures are not to be submerged while others do next to nothing. The usual height for the structures is 3–4 metres (just under 10 feet to just over 13).

Nets made of steel cable have also been used to perform the same function as snow-bridges and rakes (see Fig. 12). These nets seem best adapted to short, steep slopes above roads and railways, and their elasticity is an advantage in stopping small slides. In France, nets made of man-made fibres have been successfully used. But the main problem of all nets is in the difficulty of anchoring the cables that hold them; indeed, it is only possible to do so satisfactorily on rock. The pull of the net under snow pressure is enormous, and whole net schemes have been destroyed.

The cost of installing snow-bridges in the avalanche starting zone is enormous; not only are the structures—in steel, pre-stressed concrete or aluminium—expensive in themselves, but they have to be transported to the site and foundations prepared before they can be erected. Transportation usually means building a road or cable car, and foundations quite often have to be dug by hand. Precise costs of course vary from country to country and according to the material used. However, a rough average is about $240 per metre run (1976). This would be for steel structures 4 metres high; aluminium, installed, averages out at around 20% more. (The structures themselves are far more costly but their lightness allows savings on transport and installation costs.) Pre-stressed concrete structures, installed, usually cost about 25% less than steel.

In addition to the costs of the structures and their installation, there is frequently the high initial outlay for reafforestation, even though this can usually be considered a sound long-term investment. All in all, a defence scheme in the starting zone usually costs several million dollars before it is complete. For this reason, such schemes can only be justified to protect important installations or inhabited areas.

Most of the cost of constructing defence schemes is usually met by state subsidies. Switzerland has a particularly elaborate funding system under which the Federal Government pays 80% of the cost, the Canton 15% and the community concerned the remaining 5%. These proportions may vary slightly according to individual schemes and the ability of local interests to pay their share. Sometimes, certain commercial organizations stand to benefit from the scheme, as well as the local inhabitants, and if this is so, they are expected to pay a share of the cost. As just one example of elaborate cost-sharing, it is interesting to follow what happened in Airolo, Switzerland, in the late 1950s and early 1960s. The infamous Vallascia avalanche, which historically came close to the village almost every year, and into it occasionally, caused a larger-than-ever catastrophe in February 1951 when it leaped a 20 foot protecting wall, destroyed 29 houses and killed 10 people. (It would have killed many more had not part of the village been evacuated a few hours earlier.) A scheme to contain the Vallascia avalanche once and for all was planned and implemented at a cost of about $2·5 million. Of this the federal subsidy covered 72½% and the cantonal subsidy a further 15%. Of the remaining 12½%, the commune of Airolo paid 54%, the owners of the land affected by the avalanche 6%, Swiss Federal Railways 20%, the cantonal road authority 15%, the telephone and postal services 2½% and the Swiss Army 2½%. I have been unable to ascertain how long it took to reach such a complex arrangement, but in any case it must be something of a monument to the Swiss gift for pragmatic problem solving. It is also an example to many other countries where years of heated haggling would still not produce such an accord.

Wind Baffles. Another way of taking action against avalanches is to try to prevent the formation of cornices and large areas of wind-packed snow. This principle, unlike snow-support structures, can never be expected to give complete control, but it is useful as a complementary measure. Panels or baffles are erected where driven snow normally accumulates, and the wind flow is thereby disrupted so that only small individual slabs can form—slabs that are less likely to avalanche than a single big one. Simple wind baffles have for many years been used to cause the fall-out of wind-borne snow before it can be deposited on a road or railway, but their role in avalanche control has been pioneered more recently. The Austrians, for example, have developed an omni-directional baffle of aluminium (see Fig. 14); The Swiss and others have been experimenting with inclined boards set on top of ridges to prevent cornice and scarp creation.

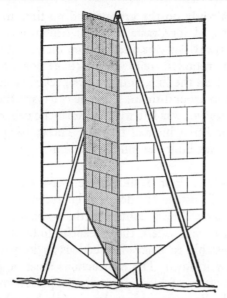

Fig. 14. The omni-directional wind baffle developed in Austria

Galleries. In the case of roads and railways, it is cheaper and simpler to protect the installation itself rather than to attempt any form of control in the avalanche break-away area. The usual method is to build a gallery over the road or railway, a strong lean-to roof that will carry the avalanche overhead. These galleries are now very common in the Alps.

The galley over the Oberalp road and railway (see photograph 38), has an interesting history. Before the Second World War the Swiss Army had formulated a plan to use the Gotthard area as a citadel in the event of invasion, and they had therefore honeycombed the mountains with tunnels and artillery emplacements. For the plan to succeed fully the Oberalp railway, which leads into the area, needed to be kept open until the invasion had actually occurred. Before the war, however, it was usual for the railway to close down for the winter, mainly because of the large avalanches that blocked the part of the line now covered by the gallery. But right up to the time war was declared nothing had been done to protect that stretch of line; and in the late autumn of 1939, a Major Crasta of the Swiss Army was told that, at all costs, the railway was to be kept running without interruption through the winter.

With snow already on the ground and no time to build a conventional gallery, Major Crasta had an ingenious idea. He obtained a number of railway flat-cars and built on them, out of sheet steel, large collapsible rounded roofs with skirtings down to the ground. He parked these on the line, and soldiers packed thousands of tons of snow over and around them. Then he collapsed the roofs, withdrew the flat-cars and was left with an avalanche gallery that served perfectly for the first winter of the war. The existing concrete and stone structure was built the following summer.

Avalanche breakers. Fair results have been obtained with yet another type of avalanche defence in recent years—so-called avalanche breakers. Where control in the break-away zone would be too expensive, and where the avalanche follows a well-defined course, it is possible to place obstructions in its path which will split it and slow it down The obstructions must be placed on the run-out at the bottom of the slope, or in a flat place somewhere in the avalanche track. Enormous mounds of earth have been used and there is a scheme incorporating these on the Hafelekar near Innsbruck. There are a few schemes using them in Switzerland, and they have also been tried in Alaska. But the Swiss have viewed them with diminished favour since an earth-mound scheme above the hamlet of Brienzwieler, Berne, allowed an avalanche to come much closer to the houses than expected. This may have been because the scheme was badly executed; but it must be realized that, in any case, such defences are only effective against avalanches that flow along the ground—against airborne-powder avalanches they are next to useless.

Other structures that have been used as avalanche breakers are massive, pre-stressed concrete tripods They are reasonably effective but early schemes involving their use were partial failures because insufficient units were installed. But, however effective avalanche breakers, splitting-wedges, deflection walls and so on may be, most of the far-sighted nations of the world that install defences prefer support structures in the break-away zone. This is, in part, owing to the fact that such schemes allow reafforestation which is regarded as an investment, whatever the intial cost.

It is not unusual for the matter of avalanche defences to raise some odd moral problems; and a good example is one in connection with the village of St Antönien in Prätigau, just north of Küblis. St. Antönien is in a backwater, one of those astonishly beautiful backwaters which the Valser people, migrating from southern Switzerland in the 13th and 14th centuries A.D., so often chose as their home. You approach the village up a narrow sinuous track among pine trees; and then, as you round a bend, the valley suddenly opens out and you see the little church, backed to the north by spectacular rock-faces Further on, you find yourself in a large sunlit bowl of grassy slopes at the base of which a little river rushes along. Neat chalets are dotted at random over the south-facing side of the bowl, and, apart from the few houses grouped near the church, there is no village centre as such. The Valser people always settled in this way, each man building his house in the centre of his land rather than forming an agglomeration.

As you look around, your first impression is one of idyllic time-lessness, of an untroubled valley in which nothing ever happens beyond the changing of the seasons, the tending of the livestock, the birth of sturdy children, and the death of adults in ripe old age after a simple and satisfying life. But a closer look belies that first impression; you notice that every house, apart from the few by the church, is backed by a massive, stone splitting-wedge. Then, on the high and wide slope above, you notice that there is but one clump of trees, a few pines set on a small protuberance. The edges of the clump are sharply defined, like those of the last patch in the centre of a cornfield when all around has been laid low. Finally, right at the head of the slope, you see row upon row of new snow-support structures, and the true nature of the place is evident. It seems almost as though you are in an amphitheatre and the stage is set for a play in which the main actors will be avalanches.

Indeed, St. Antönien has been plagued by avalanches for several hundred years. There have been no disasters of the sort that kill large numbers of people at a time: this has been prevented by the scattered nature of the village and by the steepness of the slopes, which causes the release of avalanches before they have reached very large proportions. But the regularity of the avalanches has more than compensated for any lack of size. Since 1608, avalanches have killed over 60 people in the valley. One of them, a girl, was picked up by an airborne-powder avalanche and flung over a wall into the cemetery at Christmas 1964. Also since 1608, more than 200 head of cattle, and countless smaller livestock, have been killed; and over 300 buildings and five bridges have been destroyed.

The worst single year was 1668 when 10 lives were lost; and also in that year there occurred a remarkable incident involving a young man called Conrad Ladner. He was breaking the ice at a watering place when an avalanche hit him and hurled him across the river. The river was dammed by snow and Ladner was buried on the far bank. Gradually the river undermined the mass of snow in the centre until it collapsed. A crack opened up along each bank, and one of them was directly over Ladner. He was able to climb out, unharmed, from a depth of 7 feet.

In the winter of 1807/08, 34 buildings were destroyed, even though many of them were protected by walls higher than their own roofs. In 1935, seven people were killed, and as a result of that every exposed house was provided with a splitting-wedge. In 1951, despite this, 10 people were buried, though only one died, and 42 various buildings were destroyed. When the Vorarlberg was so badly hit in 1954, St. Antönien, only a few miles away, suffered yet again. Of the four people buried two were killed, and 39 buildings were destroyed or damaged.

By 1954 the possibility of installing support structures in the avalanche break-away zones had been under study for 12 years, but the probable cost seemed hard to justify. Of the 400 inhabitants of the valley only about 150 were living in really exposed positions; and one of the early proposals was to subsidize their evacuation, or even that of the whole population. But this idea was opposed vigorously, and after the 1954 avalanches, approval was finally given for the erection of 9½ miles of support structure at an estimated cost of over $2 million. By the early 1960s more money was being called for and the final cost exceeded $3 million. Admittedly, the cost was increased by some initial errors, for this was one of the projects in which the shallow but dense snow cover of spring caused excessive loading on the lower part of the structures; many of them had to be renewed.

But even had the cost never risen above the original estimate of $2 million, one can debate whether such a sum of taxpayer's money should be spent for the protection of 150 people when they could be evacuated and installed elsewhere far cheaper. But the Swiss Government takes a long-term view: with the mountain farmer's strong attachment to his land and valley it is doubtful whether he would ever settle happily anywhere else. Another factor is that depopulation of the high Alpine valleys is already taking place to a certain extent and everything must be done to encourage the people to stay—if only because their contribution to the Swiss economy in terms of livestock produce is important. And, taking a really long-term view, it could be that villages like St. Antönien will one day grow into important tourist centres. Nor should one overlook the future dividends which will be paid by the reafforestation that accompanies such defences.

A more difficult moral problem arises in cases like that of a certain ski-resort that has been developed in the last 30 years. The original village lay in an avalanche-safe position, but when the developers and speculators moved in they built the resort on a plateau above the old site. It cannot be said that the resort is in constant danger, but about 45 years ago a reliable eye-witness did see an enormous avalanche sweep over the plateau where it now stands. Sooner or later a very large sum of money will be required to pay for an avalanche defence scheme; federal and cantonal funds will be requested to protect the installations belonging to those who were ruled by self-interest when developing the resort. The finance will doubtless be made available because even though one might like to dismiss the claims of those who developed the resort, one cannot ignore the fact that there might be hundreds of visitors in the place when an avalanche hit it.

The French have had a number of problems of this nature following the spectacular growth of skiing in France which, ostensibly, rode on the shoulders of the successes of the French national ski-team when it contained such names as Killy, Goitschel, Famose, etc. In fact, the rapid development of ski-resorts in France of recent decades was craftily engineered by de Gaulle's government when it was realized that young French athletes hurtling to triumph in race after race were creating an image of French skiing that could bring in more than medals and glory: it could also bring in tourism and foreign currency. And so new resorts were rapidly built and old resorts were expanded. One old resort was Val d'Isère. Already developed from an ancient mountain village to a modern ski-resort after the Second World War, it burgeoned with new hotels and apartment blocks in the 1960s. But what was conveniently overlooked was that the ancient village had been strategically placed plumb at the centre point of the confluence of three valleys, that is to say, as far as possible from threatening slopes. The new development took the resort towards those slopes. So no avalanche specialist was particularly surprised when, on February 10th, 1970, an avalanche ripped through a building belonging to the *Union des Centres de Plein Air* and occupied at the time by 280 youngsters, killing 39 and injuring many more.

In the general fuss that followed, Melchior Schild of the Swiss Federal Snow and Avalanche Research Institute told journalists that the French reliance on local authorities for dealing with avalanche matters was 'amateurish'. Schild has never been inclined towards tolerance, and had the journalists been more fully informed, they could have goaded him with a reminder of January 26th, 1968 when an avalanche demolished eight quite recently built houses, and damaged many others, in Davos-Dorf, another burgeoning resort, on the Institute's doorstep.

I was in Davos at the time with a British film crew making a documentary on avalanches based on the first edition of this book. A short description of some of the events of those days may help to give an insight into what it is like to be in a village under avalanche siege. It had begun to snow in the canton of Grisons and in Austria during the night of January 24th/25th when 13 cm (5 inches) of new snow fell. By the morning of the 26th, a further 51 cm (20 inches) had fallen and a fully-fledged blizzard from the north-west was in progress. We were filming in the Institute and from the time I went to look at the windgauge: 115 k.p.h. (70 m.p.h.) was about the average reading, with gusts to 140 k.p.h. (87·5 m.p.h.) showing quite often. Snow was blasting past the Institute's windows in an opaque white curtain, or in hectic swirl patterns as eddies set in.

All the ski-runs of the Parsenn were closed, so the Parsenn railway was running only often enough to keep the tracks clear. At about 16.00 hours, a Parsenn railway employee came into the Institute and told us that a special trip—and the last for that day—was to be made at 16.30. We rushed to pack up the filming equipment and, in company with the Institute staff, trundled to the valley through the raging snow-storm. That was not only the last train of the day—it turned out to be the last for several weeks.

The gravity of the situation had become obvious, and as we came down from the Weissfluhjoch in the train, we glumly looked out of the window. I suspect that the Institute staff were silently conjecturing, as I was, on what would happen and where and when, if it went on snowing with such intensity. And it did go on.

The Davos authorities were particularly concerned for the safety of the people living in the quite recently built houses in an area called 'in den Böden'. The area lies just above Davos-Dorf, on the slopes next to the Parsenn railway. It had been built on after much argument regarding its safety. During the afternoon, the mayor of Davos, in person, persuaded 300 people to evacuate from this area and from another, the Schiabach, where an avalanche in 1962 had destroyed some houses. Some families refused to move, one of them from their home in the 'in den Böden' because they were giving a party that night.

By about 18.00 hours, tension began to run high among those who knew anything at all about avalanches. I was with a group of friends in a small hotel and throughout the evening I went to the door every 20 minutes or so and looked out. There was no let-up in the snowfall and another 50 cm (20 inches) had fallen during the day.

Unbeknown to us, the first avalanche came down at about 19.00 hours just outside the town. It blocked the road and plunged into the lake. It was the Salezertobel avalanche which, as mentioned in chapter 1, had killed fish in the lake in 1569. It was the first of 12 big avalanches that came down in and around Davos on January 26th/27th, 1968.

Just after 22.30 the second came down. The first I knew of it was when an English tourist came into the room where I was sitting and hurried over to me to say that the Hotel Hermann was 'up to its eaves in snow'. Steeped in avalanche history as I was, I knew this could only mean that the enormous avalanche of 1609, the 'gruesome grisly' one described by Arduser and mentioned in chapter 1, had come down again. I went upstairs, put on the warmest clothes I could find and went out into the snow-storm to help in the rescue operation.

Only many hours later was it possible to ascertain exactly what had happened: the avalanche started high above Davos, on the flanks of the Schiahorn, plunged down fairly open slopes and then into the confines of the Dorfbach gully. It burst out of the bottom of this gully, carrying a steel bridge of the Parsenn railway with it, and then fanned out across the open slope of the 'in den Böden'.

There, it tore down a clump of mature trees, demolished eight of the recently built houses, damaged several others and finally spent its energy among the much older buildings on the uphill side of Davos-Dorf. The Hermann Hotel was one of these buildings. Until quite recently, it had been protected by a splitting wedge, in memory and respect for the 1609 avalanche. In the mid-1960s, the wedge was removed to make place for a new dining room. The avalanche of January 26th, 1968, in its death throes, wrought havoc in the dining room, but luckily the hotel guests had just vacated it.

My first impression, as I joined the local people gathering in the street close to the Hermann Hotel, was one of confusion. It was impossible to know all that I have described above, which only emerged later, and one could not see anything except the driving snow under the high street lamps; and each new arrival brought rumours about the extent of the damage and the number of missing. However, the atmosphere was one of discipline and grim calm. We awaited orders.

Within about 20 minutes, solid information began to arrive. The house with the party going on in it had not been hit, but a German couple who had refused to vacate their house were missing. Rescue equipment appeared, an official from the town council gave orders, and the 30 or so of us moved into the 'in den Böden' by the light of torches, battered by the wind and driving snow that came roaring out of the blackness at us. The avalanche had ripped the top floor off the German couple's house, leaving the lower floor, which was protected by the terrain, intact. A girl who worked as domestic help to the couple had been on the lower floor when the avalanche struck and was unharmed. But only a few minutes before, she had been on the upper floor talking to the couple. They had been watching television in the sitting room, she informed us.

By the light of our torches, we dug and sounded in the rubble-filled snow; the sense of chaos and disorientation, the feeling of helplessness in the face of the enormous area of tumbled snow and rubble, and the raging blizzard, are things I shall never forget. At 04.00 hours, cold, battered and dispirited, we were called off for fear of another avalanche down the same track.

At dawn, it was still snowing, but less heavily. Rescue operations were in course on several of the other avalanches that had come down in the Davos area; but the 'in den Böden' rescue team, of which I was part, waited all day in a nearby hotel for the weather to clear enough for a reconnaissance of the avalanche starting zone. Such a reconnaissance was needed in order to ascertain whether we could work safely or whether another avalanche might hit us. In the end, it was only the next morning that the authorities ordered us back to work.

The devastation, seen properly for the first time, was astonishing, and the task of digging through the debris seemed overwhelming. But, systematically, we began trenching at a point downslope from the remains of the house occupied by the German couple. Quite soon, one of the team struck a steel girder and called for help in freeing it. We set to, and only after several minutes did the length and section of the girder become apparent. And when they did, we suddenly realized that the girder was far too big ever to have been part of the house. A few more moments passed, however, before we recognized that the girder had been one of the main spans of the bridge of the Parsenn railway, and that it had been carried over 400 metres and hurled through the house like a battering ram. In addition, we later found short and splintered lengths of tree-trunk over a foot in diameter among the house debris.

As we dug, sawed and chopped our way through the snow and rubble, we tried to identify fragments of furniture as we came upon them, for we were on the look-out for wreckage from the sitting room where we knew the couple had been. When a piece of board with some wires hanging from it was pulled free—all that was left of the television set—we knew we were in the right area.

The heavy work was relieved from time to time by a few minutes' rest while avalanche dogs were tried; but there were too many household effects confusing the scent, so we were soon called back. We had little hope by now of finding anyone alive, but we slogged doggedly on, encouraged by the finding of the TV set. With each thrust of the shovel into the snow, I wondered what might be revealed. It was a macabre task uncovering the remnants of shattered lives—a pen, a half-written letter, a diary, a nylon stocking—all passed silently to the person making the pitiful pile of such effects. And then, at about 16.30 hours, there was a muffled exclamation as someone working a few feet from me uncovered a beautifully manicured woman's hand. 'Here!' the rescuer said loudly, as he recovered his composure. Immediately, everyone straightened from their work and a sort of sigh went up. The five or six of us nearest helped to free the body; the rest maintained silence and distance. Although not mutilated, it was clear that she had died immediately.

She was covered and loaded on to a sledge, and the supervisor of the rescue told us we could go home. But no one went. We had about 45 minutes of daylight left and the weariness in us was replaced by a frenzied need to see the horrific task finished. We attacked the snow with fresh energy, until dusk beat us and we shouldered our implements and plodded wearily down to Davos. They found the husband's body next morning.

It turned out later that another couple was also missing in the shambles of what had been new houses 'in den Böden'. Their bodies were not found for many days, bringing the death roll to 13 from the 25 people buried in and around Davos in those days of January 1968.

Such happenings have since led the Swiss authorities to a much stricter policy of zoning according to the degree of avalanche threat, only allowing building without protective measures, in the safest of mountain areas, insisting on protective measures if there is reasonable cause to doubt the zone's immunity from avalanches, and prohibiting new building completely in recognized danger areas. Until this policy was applied, it was quite common too for cable cars to be built with little heed for avalanche risk to the pylons. A case of this sort occurred at Verbier, in the Valais, a few years ago when the Tortin cable-car was built. The pylons were put up before the means for protecting them, and in the autumn the builders asked the Avalanche Research Institute for advice as to the best type of defence to construct and its necessary strength. Unfortunately the season was then too far advanced for concreting, so no splitting-wedges or other structures could be built. The only advice that could be given was to have replacement pylons prepared; these were installed in three days when the first were swept away. During the following summer, 20-foot reinforced concrete pillars were built above the pylons.

Particularly in the choice of site for a construction camp is it now usual to call for the advice of an avalanche expert. There have been a number of cases in the past of the project manager, a stranger to the area, choosing the site in summer purely from the aspect of its convenience for transport and its proximity to the work. Later, however, any convenience of this sort has been nullified through the destruc-

35. Modern aluminium snow-support structures on the Kirchberg above Andermatt. Note the saplings that have been planted. Photograph taken after a few inches of snow had fallen in early autumn

36. Aluminium support structure on the slope near the triangular wood above Andermatt. The Kirchberg is out of the picture in the top right-hand corner, above the barracks

37. General view of support-structure scheme near Andermatt

38. Gallery over the Oberalp road and railway (see p. 233)

tion of the camp by an avalanche. But, even after an expert's opinion has been sought, the same thing may happen. Advice was called for when placing a hostel for 300 workmen engaged on the Italian section of the Mont Blanc tunnel; yet, in April, 1962, the building was destroyed by an avalanche with the loss of three lives and 35 other men injured.

THE AVALANCHE-WARNING SYSTEM

Avalanche-warning systems have been developed in many parts of the world in the last 25 years. Not that avalanche forecasting is new: a book of 1829, entitled *The Avalanche, or the Old Man of the Alps* and written by a Madame Montolieu, recounts the story of a 100-year-old shepherd in the Alps who predicted disastrous ice avalanches from his observations of the glacier where they originated. He was not always believed, however—and nor are all avalanche warnings even today. So, on one occasion during the last year of his life and when, according to his judgement, an avalanche threatened a hamlet, he and a young engaged couple pulled a ruse by arranging for the date of their wedding, which was to take place in a nearby village, to be fixed on the day he estimated as the most probable for the avalanche to happen. With nice timing, the avalanche came down on to the deserted hamlet, and the old man was the hero of the moment among the inhabitants carousing safely at the wedding in the unharmed village. The old man, in his response to the acclaim, said that his modest observations were merely the forerunners of the scientific research that would take place one day.

The first organized, albeit primitive, warning system came into existence in the Alps in the winter of 1937/38 when the Swiss Ski Club issued press and radio bulletins on 18 successive Fridays throughout the season. The bulletins were made up as the result of observations taken in 15 different areas, and it was possible to give a general guide to the level of avalanche danger which skiers would face over the weekend, assuming no radical change in conditions.

When the Second World War broke out the Swiss Army realized that such bulletins could be invaluable for the safe movement of troops in the mountains, so the Avalanche Research Institute at the Weissfluhjoch was given the task of expanding and improving the service. By the end of the war there were 25 measuring stations sending in weekly reports to the Weissfluhjoch from where a bulletin was issued every Friday. In the event of particular danger, extra reports could be sent in and special bulletins issued. The system functioned satisfactorily until February 1951. The avalanche conditions produced by the January storms in northern Switzerland had been correctly analysed, and the warning bulletins were accurate in nearly every respect; but the February storms south of the Gotthard showed up a grave weakness in the organization. The storm brought down the telephone lines in the neighbourhood of Airolo and that part of the Tessin was completely cut off from the Avalanche Research Institute. No one at the Weissfluhjoch had any idea that such an extraordinary storm was raging south of the Gotthard.

Thereafter, it was decided to increase the number of measuring stations in Switzerland to 50, which would allow some sort of picture to be built up by interpolation should the odd station not be able to report; and the communication system was also changed. The telephone is now only used to ring the nearest telegraph office—a journey the observer could make on skis if necessary—and from there his report goes to Zurich Airport to be transmitted by teleprinter to the Avalanche Research Institute.

The observers who take the readings are a mixed collection of volunteers whose normal job may be anything from border guard to schoolmaster. They attend training courses at the Avalanche Research Institute and then, for a very meagre monthly payment while snow is on the ground, they go out to their measuring station each morning and make a series of observations. From the first snow-fall to the moment when the earth is bare again (but beginning at the latest on December 1st and continuing until at least April 30th) they must make their daily reports. By about 09.30 each day the last of these reports has clattered over the Institute teleprinter.

The report from each of the 50 measuring stations comes through as a line of 42 figures, divided into 10 groups, and these figures give a very complete picture of the weather, snow and avalanche conditions at the station. The following is the actual report from Andermatt for a January 1st morning:

02an 0101 0815 0103 2301 95400 01552 21209 80000 71373

In the first group the 02 signifies the region (Switzerland is divided into seven climatic regions for the purpose of these reports), and the 'an' stands for Andermatt. The second group gives the day and month, in this case the 0101 of course means first day of first month. The 0815 in the third group is the time of the observation—which is normally before 0800, but to be only quarter of an hour late on New Year's morning seems to me quite praiseworthy. In the fourth group of figures are given the general weather conditions and cloud cover. The first two figures of the group can be anything between 01 and 95, and every possible eventuality is covered from, for example, 'fog but with sky visible' to 'precipitation of fine ice needles', or even 'occasional fine rain with icing'. In the case of our Andermatt report, the 01 means 'cloud decreasing'. The next two figures in the group give the cloud cover in tenths, the 03 here of course meaning three-tenths. In the fifth group is the wind direction in compass degrees and the strength of the wind in Beaufort Scale. Here the 23 indicates 230°, and the 01 means Force 1—that is to say that there was a light, south-westerly air.

The sixth group begins with the invariable prefix 9 which shows that the report is going on to deal with temperature and snow conditions. The first two figures after the 9 are the air temperature in Centigrade and, to distinguish temperatures below zero, 50 is always added to them; thus the 54 here means −4°C. The next two figures are the new snow in centimetres, in this case none. In the seventh group the first three figures give the total snow depth in centimetres. (Not so many years ago, only 15 cm of snow in Andermatt on January 1st would have been extraordinary; Alpine snowfall has become erratic in the last 10–15 years and *seems* to be moving more and more towards what used to be spring.) The last two figures in the same group are the temperature of the snow 10 cm below the surface, in this case (using the rule of adding 50 to minus figures) −2°C. The first two figures of the eighth group can be anything between 01 and 37 to denote the characteristics of the snow surface, for instance, 'wind-slab', 'firm as the result of settling', 'loose', etc. In the Andermatt report, the 21 signifies 'feathery snow'. The next figure, which can be from 1 to 6, indicates the form of the snow surface, that is to say whether it is 'flat', 'eroded', 'surface hoar' and so on. The figure 2 in our report means 'wavy'. The last two figures in the group, here 09, is the depth in centimetres of a ski-track.

The ninth and tenth groups, beginning with the invariable prefixes 8 and 7, directly concern avalanches. In the ninth group are details of avalanches that have been seen. In this report, the four zeros indicate that there have been none; but in place of the first zero there could be a figure from 1 to 6 to describe the type of avalanche; in place of the second zero a figure up to 9 would give the type and orientation of the slope; in place of the third zero there would be a figure to indicate the altitude of the break-away point; and in place of the last zero a figure would denote the number and size of avalanches seen.

In the tenth group, with the prefix 7, the observer gives his assessment of the avalanche danger. In the Andermatt report, the 1 indicates a danger of dry-slab avalanches; the 3 that they will occur on mainly shaded slopes; the 7 that the break-away zone will be above 2,500 metres; and the last figure denotes the degree of danger —the 3 in this case means 'moderate and constant'. Had there been any new snow the report would have included some more figures following a prefix 6. They give the weight of new snow held in a tube of given volume and thereby allow the snow density to be calculated.

As mentioned earlier, several of Switzerland's neighbouring countries now have avalanche-warning systems, and they all use the same numerical coding system for observers' reports. Those who receive these coded observations and are responsible for drawing up warning bulletins can gain a concise idea of the prevailing conditions in the areas they cover. In addition, they also have fortnightly snow profiles from the same stations to tell them what is going on deep in the snow cover. But the interpretation of the data into an assessment of where danger may lie—at what altitudes and on what orientation of slopes—and how great that danger is, requires not only knowledge of the ways of snow and avalanches, but also a fair dash of intuition based on experience.

I drew up the bulletins for the Central Appennines of Italy for four winters in the early 1970s, before the national Forestry Corps personnel who worked with me on them felt sufficiently experienced to carry on alone. I had often sat in with Melchior Schild of the Swiss Institute, wreathed in his acrid cigar smoke and listening to him thinking aloud as he analysed the available data and contemplated especially important features such as new snow-fall, strong winds or rapid temperature changes. His assessment resulted in a draft bulletin which he would take to a meeting with Dr. de

Quervain and some other Institute staff for comment and finalization. As a matter of course, the bulletin was, and still is, released every Friday to press, radio and television, but extra bulletins are put out at other times if the situation changes significantly for worse or for better. This practice has been copied by most countries.

When I came myself to take on this task of issuing weekly, or more frequent, avalanche bulletins covering recent snow and weather conditions, the degree of avalanche danger, the types of avalanche most likely to occur, and where, I discovered what an agonizing experience it can be. There is a natural tendency to play for safety, to cover oneself against the fool who gets himself killed during some suicidal venture into a steep couloir where very localized danger may have persisted for many weeks, danger which one could hardly have predicted in a more generalized assessment. And yet, if one consistently exaggerates the danger, no one believes the bulletins any more. In the end, the compromise solution I adopted was to be as objective as possible in presenting the danger as I assessed it, but when I felt that some menace might still be lurking in odd nooks and crannies, I wound up the bulletin with a few general words of caution about localized danger to ski-alpinists undertaking tours.

All in all, however, even though no one was ever caught in an avalanche when I had issued a bulletin announcing negligible danger, I was always aware of the degree of subjectivity that inevitably creeps into assessments drawn from lines of figures interpreted according to personal experience and observation, and tinged too by the desire to err on the cautious side. Therefore I feel that although the existing avalanche-warning systems in Europe have generally given excellent results, there is too much margin for human error for them to be ideal. And you only have to listen to compilers of avalanche bulletins criticizing their colleagues' bulletins to realize the degree of variation there can be in the type of bulletin that emerges from the same data. Another drawback is that, as warning systems of this type are expanded to new regions or countries, more and more top avalanche expertise is required if the systems are to work properly; and a good compiler of bulletins cannot be trained from one winter to the next.

In the United States, there has, for a long time, been more effort directed towards establishing criteria by which to judge an avalanche situation and produce an appropriate bulletin. This does not mean that American avalanche experts produce their bulletins by rule-of-thumb methods, without drawing on their experience; but for all the factors that their European counterparts take into *general* consideration the Americans have tried to establish certain critical levels. For example, it is well known that the intensity of a snow-fall has a strong bearing on avalanche formation, and in the United States a so-called Precipitation Intensity Factor has been introduced. The precipitation intensity in this context is the rate at which water, in the form of snow, is deposited by snow-fall and/or wind drift, and it is held that if a storm deposits water at a rate greater than 0·10 inch per hour, and continues to do so for more than 10 hours, an avalanche situation will have been reached.

And there are several other standards used in the United States. Experts there maintain that wind begins to influence avalanche formation at speeds in excess of 15 m.p.h., and that by 30 m.p.h. it plays a very important part indeed. They hold that a settlement rate in new snow of more than 20% in 24 hours brings about a considerable reduction in avalanche hazard. They take the normal density of new snow as being 0·07–0·10 grams/cc, and they state that the wider the departure from this range, in either direction, the greater the avalanche danger. (At the extremes, low density brings the risk of 'wild' and loose-snow avalanches, and high density the risk of hard-slab avalanches.) And they state that, in average mountain terrain, 2–3 feet of old snow will be sufficient to level out irregularities and provide a smooth base on which further snow-falls may avalanche. Overall, the American experts take a list of 10 factors into consideration when preparing an avalanche bulletin: old snow depth, condition of base, new snow depth, new snow type, new snow density, snow-fall intensity, precipitation intensity, settlement rate, wind and temperature. Of course, in a given set of circumstances, the effect of many of these factors may be contradictory; so it is suggested that, in the analysis by the expert, each be given a mark from 0–10 for the degree of avalanche danger it is contributing—the higher the danger the higher the mark. If, out of the possible 100 marks, the score is in the 40–60 range the danger is marginal, while any score over 75 would indicate a definite impending hazard.

This system is only valid for warning of avalanches that take place during or immediately after a storm. For 'delayed action' slab avalanches, which may occur at any time if a fragile stratum in the snow cover is overloaded to breaking point, an American once suggested that a slab with a penetrometer resistance of 100 kilograms lying on a stratum with a resistance of only 10 kilograms constituted a threat. Undoubtedly this is true, but it is too arbitrary to be used as a standard because slab avalanches also occur when the figures differ very greatly from those quoted.

Even though European avalanche experts take into account the same general things when making out an avalanche bulletin, they refuse to be bound by any set standards as in the United States. I believe that the European reluctance in this respect stems from a greater awareness of the fact that, despite modern research, avalanches still have an unpredictable streak. They have done the unexpected too often in Europe; too many 400–500-year-old buildings have been destroyed. After all, if an avalanche comes down a given slope in the United States no one knows whether or not it is for the first time in 500 years, because there can be no house that old at the bottom for it to destroy.

Unfortunately, I have no personal experience of the American Avalanche-Warning System, but from all reports it gives good results. I would not presume to say, as do many Europeans, that it is too cut and dried. On the contrary, I believe that the greatest future lies in the recent initiatives, mentioned earlier in this book, to develop mathematical models of types of avalanche activity as a function of all the variables that go into avalanche creation. It will take many years of precise observation and analysis to draw up these models, but initial and very preliminary results from the work carried out in France and Switzerland do show promise. I feel it will indeed be possible therefore, one day in the future, to feed multi-variate data into a computer and obtain a calculated assessment of the actual danger. No doubt, the models will need almost continual adjustment in the light of experience over the years before the ultimate models are established. And even then, avalanches may still do the unpredicted occasionally—I would be rather disappointed if they did not —but such a system must ultimately give better results than the systems in general use today.

*　　*　　*

Despite all the effort and resources that go into eliminating avalanche danger to mankind, more and more people are exposed to it every year. This is because of the continual growth in the popularity of skiing, a sport which is already said to have more active participants than any other in the world.

It is hardly surprising, then, that it becomes easier and easier for avalanches to find victims, most of them unsuspecting and tending anyway to underestimate the menace. I am reminded of an incident that occurred about half an hour after we had rescued Georg Caviezel from the avalanche that buried him on the blasting operation described in chapter 10. Heini Schwendener and I were standing by the Kreuzweg hut, the point at which the Küblis and Klosters ski-runs divide. Christian Jost had told us to wait there so that we could go to the help of some colleagues, who were blasting Gauder-grat on the Küblis run, in the event of an accident. When they had finished we were to blast some slopes on the way to Klosters.

While we were waiting, two skiers came down and stopped close by us. They were English and began to debate which run to take. To reach us they had passed two avalanche danger signs and two signs informing them that the run was closed. They had also passed all the avalanches that we had just brought down with explosives; and they had even skied over the debris of the one that had crossed the run.

After a few moments discussion one said, with a nod in our direction:

'Let's ask them.'

'No,' said his friend scornfully. 'They'll only tell us not to go.' And they began to move off.

'I wouldn't go.' I said. 'And don't you know this is closed down here?' They turned, obviously astonished at hearing a Parsenndienst man speak fluent English, and also somewhat embarrassed that I had overheard their remark.

'Oh yes, we know it's closed, but why is it?' one of them asked. Now, when avalanches are in evidence all around, the usual reply given to that question by the more humorous patrolmen, is that there is great danger of landslides and rock-falls! But I was still somewhat shaken by Caviezel's accident and was not in a mood for joking.

'Because of avalanches,' I said.

'Oh, it's not as dangerous as all that,' said one of the skiers.

His reply exhausted my patience and I gave vent to my feelings in some straight talking, ending with the words: 'And if you don't think it's dangerous you should bloody well have been with us half an hour ago when we were lucky to rescue a colleague alive.'

There was really no valid excuse for my lack of courtesy and I have since come to regret it; for there are two sides to every story, and I am now convinced that at the root of most of the troubles in ski areas is a lack of understanding between tourists and the rescue and safety organizations. In truth, one of the very few things that I can reproach the Parsenndienst with are its poor public relations. And, judging from other places I have skied in, from hundreds of skiers I have spoken to, the same weakness seems to be general. Visitors often have not the vaguest idea of what goes on behind the scenes in a ski-area, or of the problems that face the organizations responsible for their safety. It is one of my hopes that this book will do just a little to alter this situation.

A good example of this lack of communication took place in Davos one February. Some of the runs were closed after a series of light snow-falls—light enough for few people to realize that localized avalanche danger had been created, owing to the wind that accompanied the snow. The visibility then remained so bad that the Parsenndienst were unable to go out blasting, and the runs stayed closed for three days.

On the third evening, an irate Englishman asked me why some runs were not open. He complained at length that there had not been enough snow to create avalanche hazard; that one of them was his favourite run; that he paid good money to ski at Davos; and that the Parsenndienst were inefficient. In the end I managed to explain that localized avalanche danger did indeed exist, and that, owing to the fact that the visibility had not exceeded 50 yards for more than five minutes at a time during the past three days, it had been impossible to go blasting. I told him of the dangers of blasting, both to tourists and patrolmen, if one could not see properly.

He was at once placated and reasonable; but he asked why this sort of information was not made available to the public, adding that most people's cooperation would be secured if it were. He then told me that the strangest rumours were circulating to explain the lengthy closure of those runs. One was that the hospital had no more room for skiers with broken legs; and another was that a skier had broken his leg four days before on one of the runs and that, in the fog, the Parsenndienst had not been able to find him before he died. However ill-founded and ridiculous such rumours may be, they are bound to be damaging.

I am sure that to keep skiers informed would in fact make them more cooperative. The Englishman who made that point told me that he had been very tempted to ski down his favourite run because, by the third day, he could see no reason for its closure. Many organizations like the Parsenndienst do issue little leaflets in three languages outlining their activities, but this is not enough. Explanatory notices should be placed beside the 'Run Closed' sign, at least in the main cable-car and funicular stations, if the reason for the closure is not obvious. I think that in the case just cited a notice should have been displayed which read something like the following: 'We much regret the continued closure of such-and-such ski-runs, but they are still menaced by localized avalanche danger. Owing to the bad visibility of the last three days it has not been possible to make the runs safe with explosives, though this will be done as soon as possible.'

In the U.S., skiers are often encouraged to watch from a safe place while avalanche control with gunfire is in progress. At the same time, a snow ranger explains what is being done and why. As a means of educating the public and gaining their help this must be the right approach, though this particular practice could not be used in Europe because of language difficulties: it would not be vital if some of the skiers present did not understand the explanation— but a dangerous situation could result if someone did not understand that he had to stay in a safe place while the blasting was in progress.

But, even with good communications between tourists and safety organizations, there will always be the foolhardy who choose to ignore advice out of bravado. I suspect that the two Englishmen who arrived at the Kreuzweg hut that morning fell into this category. They were quite good skiers and they *must* have recognized that there was avalanche danger. I always think that the words of Emil Zsigmondy in his book *Die Gefahren die Alpen* (The Dangers of the Alps) are very appropriate in the face of such behaviour: 'It is fine to prove one's audacity and intrepidity in the accomplishment of a task which others consider impossible; but no one will admire the hero if he dies in the attempt.' And, as has been explained earlier, it is not only themselves that such people endanger when they ski on closed runs: they leave tracks that may lure other skiers to their deaths, skiers with no desire to be either audacious or intrepid.

Complete protection against those avalanches that terrorize the population of the mountains in winters like that of 1950/51 will be a long time in coming—if it ever comes. The size of the problem will be realized from the fact that in January 1951 there was an average of four avalanches for every mile of valley in many parts of Switzerland. And in some valleys the density was far greater, as at Davos-Monstein where 15 avalanches came down in a 2-mile stretch. To build sufficient snow-support structures to prevent an occurrence of this magnitude is, of course, out of the question in the foreseeable future. And, forgetting the Alps for a moment, a number of mountainous areas of the world are undergoing development. Avalanches are common in parts of Latin America, in Alaska, in many countries of the Near East, and in some parts of Asia, and Australasia; every new road, and every new settlement, may provide a target where avalanches have gone down harmlessly in the past.

Avalanches, then, are great destructive forces, spectacular phenomena that can inspire admiration and awe, and their enigmatic vagaries hold an alluring fascination. It seems probable that they will menace the people of the mountains, and those who go there in search of pleasure, for a very long time to come.

Bibliography

The following list, though not exhaustive, is a broad selection from the existing works on snow and avalanches and from works that contain interesting references to the latter.

ALLIX, A. 'Les Avalanches.' *Revue de Geographie Alpine* (Grenoble), vol. 13 (1925), pp. 359–424.

ALLIX, A. 'Les avalanches de 1922–23 en Dauphiné.' *Revue de Geographie Alpine* (Grenoble), vol. 11 (1923), pp. 513–527.

ALLIX, A. 'Les premiers textes dauphinois relatifs aux avalanches. *La Montagne*, vol. 21 (1925), pp. 44–49.

ATWATER, M. 'A study of debris distribution in a major avalanche.' *Appalachia*, vol. 29 (1952).

ATWATER, M., LA CHAPPELLE, E., and others: 'Avalanche Research— a progress report, part 1.' *Appalachia*, vol. 30 (1954/55), pp. 209–220.

ATWATER, M., LA CHAPPELLE, E., and others: 'Avalanche Research—a progress report, part 2.' *Appalachia*, vol. 30 (1954/55), pp. 368–380.

ATWATER, M., KOZIOL, F. C. *The Altah Avalanche Studies* (U.S. Dept. of Agric. Forest Service, 1950).

BÄCHTOLD, A. 'Staublawinen.' *Die Alpen*, vol. 13 (1937), No. 1, pp. 5–7.

BENTLEY, W. A. HUMPHREYS, W. J. *Snow Crystals* (New York & London, and New York 1931). 1962.

BOIS, PH., CH. OBLED, W. GOOD. *Multivariate data analysis as a tool for day by day avalanche forecast.* (Proc. Int. Symp. on Snow Mech., Grindelvold 1974)

BUCHER, E. 'Beitrag zu den theoretischen Grundlagen des Lawinen-verbaues.' *Geologie der Schweiz—Geotechnische Serie—Hydrologie, Lieferung* 6 (Bern, 1948).

BUCHER, E., JOST, CHR. 'Expériences de déclenchement artificiel d'ava-lanches au moyen du lance-mine.' *Neue Zürcher Zeitung* (31/7 & 1/8/1941).

BUCHER, E. 'Technische Überlegungen zum Problem der Lawinenbil-dung.' *Berge der Welt* (Zürich, 1946).

BURTON, R. G. 'Napoleon's Campaign in Italy' (London, 1912).

BUSS, E. 'Ueber die Lawinen.' *Jahrbuch Schweiz. Alpenclub*, vol. 45 (1910), pp. 250–273.

BÜTLER, M. 'Der Luftdruck bei Staublawinen.' *Die Alpen*, vol. 13 (1937), pp. 11–12.

BÜTLER, M. 'Stärke und Geschwindigkeit des Lawinenluftdruckes.' *Schweiz. Zeitschrift für Forstwesen*, vol. 89 (1938), No. 4/5.

CAMPBELL, R. 'Bergfahrer und Lawine.' *Die Alpen*, vol. 10 (1934).

CECIL ALTER, J. 'Why the snow slides from the mountain slopes.' *Monthly Weather Review*, vol. 40 (1912), pp. 608–609.

CHURCH, J. E. 'Snow Perils and Avalanches.' *The Scientific Monthly*, vol. 56 (1943), pp. 309–331.

COAZ, J. *Der Lawinenschaden im schweiz. Hochgebirge im Winter und Frühjahr 1887/88* (Bern, 1889).

COAZ, J. *Die Lawinen der Schweizeralpen* (Bern, 1881).

COAZ, J. *Statistik und Verbau der Lawinen in den Schweizeralpen* (Bern, 1910).

COOLIDGE, W. A. B. *Alpine Studies* (London, 1912).

COOLIDGE, W. A. B. *The Alps in Nature and History* (London, 1908).

CORREVAN, E. 'Une Avalanche meutrière.' *Die Alpen*, vol. 31 (1955), No. 2 (after article from *Echo des Alpes*, 1914).

CLARK, R. *Great Moments of Rescue* (London, 1951).

CLARK, R. *The Early Alpine Guides* (London, 1949).

DE BEER, Sir G. *Alps and Men* (London, 1932).

DE BEER, Sir G. *Early Travellers in the Alps* (London, 1930).

DE BEER, Sir G. *Travellers in Switzerland* (London, 1949).

DE CAYEUX, A. 'Les Avalanches.' *Geographia* (Paris), vol. 41 (1955), pp. 28–31.

DE QUERVAIN, M. 'Die Entwickling der Schweiz. Schnee und Lawinenforschung von 1942 bis zur Gegenwart.' *Schweiz. Zeitschrift für Forstwesen*, No. 12, December 1961.

DE QUERVAIN, M. 'On the metamorphism of snow.' Chapter in *Ice and Snow* (M.I.T. Press, 1962).

DE QUERVAIN, M., DE CRECY, L., LA CHAPELLE, E. R., LOSEV, K., SHODA, M., *Proposal of the working group on avalanche classification of Int. Comm. of Snow and Ice*. (Hydrol. Sci. Bull. 18 (4) pp. 391–402

DE QUERVAIN, M. 'Von der Arbeit der schweiz. Schnee und Lawinenforschung.' *Die Alpen*, vol. 31 (1955), No. 2.

DUMARTHERAY, J. 'Avalanches!' *Die Alpen*, vol. 13 (1937), pp. 117–120.

DUMAS, COUNT MATTHIEU. *Memoirs of his own Time* (London, 1939).

FRANKHAUSER, F. 'Ueber Lawinen und Lawinenverbau.' *Die Alpen*, vol. 5 (1919), No. 1.

FAVRE, B. 'Une avalanche imprévisible.' *Die Alpen*, vol. 31 (1955), No. 2.

FLAIG, W. *Lawinen!—Abenteuer und Erfahrung, Erlebnis und Lehre* (1st ed., Leipzig, 1935; 2nd ed., Wiesbaden, 1955).

FLAIG, W. *Lawinen Franz-Joseph* (Munich, 1941).

FORBES, J. D. *Travels through the Alps of Savoy* (Edinburgh, 1843).

FOREL, F. A. 'La'avalanche du glacier des Têtes Rousses.' *Comptes Rendus*, CXV (1892), pp. 193–196.

FRONTIER BOOIS. *The incredible Rogers Pass* (Book 8, Frontiers Unlimited, Calgary).

GALLICIOTTI, F. *Il flagello bianco nel Ticino* (Bellinzona, 1954).

GERBER, E. 'Staublawinen.' *Die Alpen*, vol. 13 (1937), pp. 9–10.

GODEFROY, R. 'Les Avalanches, d'après un ouvrage récent. *La Montagne*, vol. 22 (1926), pp. 187–195.

GOS, C. *Alpine Tragedy* (London, 1948).

GOSSET, P. C. 'Narrative of the fatal accident on the Haut-de-Cry.' *Alpine Journal*, vol. 1 (1870), pp. 288–294.

GROB, W. 'Some thoughts about avalanches.' *Sierra Club Bulletin* (San Francisco), vol. 35 (1950) p. 91.

HAEFELI, R., BADER, H., BUCHER, E., NEHER, J., ECKEL, O., THAMS, CHR. 'Der Schnee und seine Metamorphose.' *Geologie der Schweiz—Geotechnische Serie—Hydrologie, Lieferung 3* (Bern, 1939).

HAEFELI, R., DE QUERVAIN, M. 'Gedanken und Anregungen zur Benennung und Einteilung von Lawinen.' *Die Alpen*, vol. 31 (1955), No. 2.

HAEFELI, R., BUCHER, E. 'Recherches récentes en matière de lutte contre les avalanches.' *L'Annuaire de l'Association Suisse de Clubs de Ski*, vol. 35 (1939).

HAEFELI, R. 'Stress transformations, tensile strengths, and rupture processes of the snow cover.' Chapter in *Ice and Snow* (M.I.T. Press, 1962).

HAEFELI, R. 'Tätigkeitsbericht 1934 bis 1937 der Schweiz. Kommission für Schnee- und Lawinenforschung, Station Davos-Weissfluhjoch.' *Schweiz. Bauzeitung*, vol. 110 (1937), No. 8.

HAEFELI, R. 'Von den Angfangen der Schnee- und Lawinenforschung.' *Schweiz. Zeitschrift fur Forstwesen*, No. 12, December 1961.

HAEFELI, R. 'Zur Beobachtung der winterlichen Schneeverhältnisse in den Schweizer Alpen. *Die Alpen*, vol. 21 (1945), No. 3.

HAEFELI, R. 'Zur Entwicklung der Schnee- und Gletscherforschung. *Wasser- und Energiewirtschaft* (1960), No. 8/10.

HAEFELI, R. 'Zur Geschichte der Bremsverbauung von Lawinen.' *Schweiz. Zeitschrift für Forstwesen*, No. 8, August 1960.

HENCHOZ, L. 'La'avalanche.' *Die Alpen*, vol. 31 (1955), No. 2.

HERRLIBERGER, D. *Topographie der Schweiz.* (Basel, 1773).

HESS, E. 'Schneebrettlawinen.' *Die Alpen*, vol. 10 (1934), pp. 81–95.

HESS, E. 'Schneeprofile.' *Jahrbuch des Schweiz. Skiverbandes*, 1933.

HESS, E. 'Wildschneelawinen.' *Die Alpen*, vol. 7 (1931), pp. 321–334.

HOEK, H. 'On Snow Avalanches.' *Alpine Journal*, vol. 23 (1907), pp. 379–386.

INTERNATIONAL 'VANNI EIGENMANN' FOUNDATION, MILAN. 'Symposium über dringliche Massnahmen zur Rettung von Lawinenverschütteten mit Berücksichtigung der wissentschaftlich-technischen Hilfsmittel.' (Collected papers presented by various authors at symposium held at Davos, January 1963.)

KANT, I. *Immanual Kants Physische Geographie* (Hamburg, 1817).

KOEGEL, L. 'Schnee und Lawinen.' *Geographische Zeitschrift* (Leipzig & Berlin), vol. 46 (1940), pp. 132–138.

KOHL, J. G. *Alpenreisen*, vol. 3, 'Die Lawinen' (Dresden & Leipzig, 1851).

KOPP, M. 'Maturation de risque d'avalanche.' *Jeunesse forte-peuple libre*, February 1962.

KRASSER, L. *Grundzüge der Schnee und Lawinenkunde* (Bregenz, 1964).

KRAUS, E. 'Ueber Grundlawinen.' *Zeitschrift Deutschen Geol. Gesellschaft*, vol. 90 (1938), No. 4.

KUGY, J. *Alpine Pilgrimage* (London, 1917).

LA CHAPELLE, E. R. *Field Guide to Snow Crystals*. (Seattle, 1969)

LORETAN, R. *Die Lawinenverbauungen Torrentalp-Leukerbad*. (Inspektion für Forstwesen, Bern, 1935.)

LUNN, Sir A. 'Schilthorn Avalanche.' *British Ski Year Book*, 1925.

LUNN, Sir A. *Skiing at all Heights and Seasons* (London, 1921).

LUNN, Sir A. *The Story of Skiing* (London, 1952).

MARINER, W. *Neuzeitliche Bergrettungstecknik—ein Leitfaden für den Ausbildung des Rettungsmannes* (Innsbruck, 1949).

MATHEWS, C. E. *The Annals of Mount Blanc* (London, 1898).

MONTOLIEU, Mme. *The Avalanche, or the Old Man of the Alps* (London, 1829).

MORELL. *Scientific Guide to Switzerland* (London, 1867).

NAKAYA, U. *Snow Crystals* (Cambridge, Mass., 1954).

NOELTY, F. A. M. 'Concerning avalanches and security therefrom. *Alpine Ski Club Annual*, 1912.

OECHSLIN, M. 'Das Spiel der Lawinen.' *Die Alpen*, vol. 31 (1955), No. 2.

OECHSLIN, M. 'Der Kampf gegen die Lawinen.' *Die Alpen*, vol. 31 (1955), No. 2.

OECHSLIN, M. 'Lawinengeschwindigkeiten und Lawinenluftdruck.' *Schweiz. Zeitschrift für Forstwesen*, 1938, No. 1.

OECHSLIN, M. 'Schneetemperaturen, Schneekriechen und Kohäsion.' *Schweiz. Zeitschrift für Forstwesen*, 1937, No. 1.

OECHSLIN, M. 'Wie unsere Altvordern die Lawinen beschrieben.' *Die Alpen*, vol. 31 (1955), No. 2.

OERTEL, E. *Die Lawinengefahr und wie der Alpinist ihr begegnet* (Munich, 1925).

PAULCKE, W. *Lawinengefahr, ihre Entstehung und Vermeidung* (Munich, 1926).

PAULCKE, W. 'Schneewächten und Lawinen, Ergebnisse meiner Schnee-forschung. *Zeitschrift des Deutsch. und Österr. Alpenvereins*, 1934, p. 247.

PAULCKE, W. *Praktische Schnee und Lawinenkunde* (Berlin, 1938).

PUTNAM, W. L. 'Snow Conditions 1—The individual crystal and its evolution.' *Appalachia*, vol. 28 (1950/51), pp. 56–61.

PUTNAM, W. L. 'Snow Conditions 2—Fresh and powder snow avalanches.' *Ibid.*, pp. 171–175.

PUTNAM, W. L. 'Snow Conditions 3—Windslabs.' *Ibid.*, pp. 393–397.

PUTNAM, W. L. 'Snow Conditions 4—Wet snow and névé.' *Ibid.*, pp. 562–565.

PUTNAM, W. L. 'Snow Conditions 5—How to start an avalanche.' *Appalachia*, vol. 29 (1952/53), p. 45.

RICKMERS RICKMERS, W. 'At random—two avalanches.' *Alpine Ski Club Annual*, 1908.

RICKMERS RICKMERS, W. 'The Avalanche.' *Alpine Ski Club Annual*, 1912.

RIDDELL, J. 'Six minutes too soon.' *British Ski Year Book*, 1952.

ROCH, A. 'An approach to the mechanism of avalanche release.' *Alpine Journal*, vol. 70 (1965), No. 310.

ROCH, A. 'Avalanches.' *Mountain World* 1962–63, pp. 28–39 (Zürich, 1964).

ROCH, A. 'Avalanche Danger in Iran.' *Journal of Glaciology*, vol. 3, No. 30, October 1961.

ROCH, A. 'Le méchanisme du déclenchement des avalanches.' *Die Alpen* vol. 31 (1955), No. 2.

ROCH, A. 'L'espacement des râteliers de retenue de la neige.' *Schweiz. Bauzeitung*, vol. 73 (1955).

ROCH, A. 'L'Institut Fédérale pour l'Etude de la Neige et des Avalanches.' *Le Ski*, vol. 11 (1962), pp. 413–417.

ROCH, A. 'On the study of avalanches.' *Sierra Club Bulletin* (San Francisco), vol. 63 (1951), No. 5, pp. 88–93.

ROCH, A. 'Pourquoi la neige a-t-elle des qualités extrêmement variées?' *La Route et la Circulation Routière*, vol. 44 (1958), No. 4.

ROHRER, E. 'Luftbewegung bei Staublawinen.' *Die Alpen*, vol. 31 (1955), No. 2.

ROHRER, E. 'Staublawinen—ja oder nein.' *Die Alpen*, vol. 30 (1954), No. 3, pp. 62–68.

ROUSSET. *Mémoires du Maréchal MacDonald* (Paris, 1848).

SCHEUCHZER, J. J. *Beschreibung der Natur-Geschichten des Schweizerlandes* (Zürich, 1706).

SCHILD, M. 'Lawinenhünde.' *Schweizer Spiegel*, February 1959.

SCHILD, M. 'Skifahrer, Vorsicht: Lawine!' *Gewerbschuler Leseheft*, 41/4.

SCHILD, M. 'Über die Entstehung des Lawinenbulletins.' *Die Alpen*, vol. 31 (1955), No. 2.

SCHLUMPF, A. 'Zur Mechanik der Staublawine.' *Die Alpen*, vol. 13 (1937), pp. 85–94.

SCHMIDKUNZ, W. *Der Kampf über den Gletschern* (Munich, 1918).

SELIGMAN, G. 'Comments on avalanche research.' *Journal of Glaciology*, vol. 2, April 1953, p. 233.

SELIGMAN, G. 'Les Avalanches.' *Sports d'Hiver* (Paris), No. 44, January 1937.

SELIGMAN, G. 'Snowcraft and avalanches for the skier.' *Newsletter, Chicago Mountaineering Club*, vol. 13 (1959), pp. 2–9.

SELIGMAN, G. *Snow Structures and Skie Fields* (London, 1936).

SELIGMAN, G. 'Snow Structures—some practical applications.' *Journal Royal Met. Soc.*, vol. 63 (1937), pp. 93–103.

SELIGMAN, G. 'Wind Slab Avalanches.' *Journal of Glaciology*, July 1947, pp. 70–73.

SENNET, R. *Across the Great St. Bernard* (London, 1904).

SIMMLER, J. *Commentarius de Alpebus* (1574; translation into French by Coolidge, 1904).

SMITH, ALBERT. *The Story of Mont Blanc* (London, 1853).

SMYTHE, F. S. 'Some physical characteristics of snow avalanches.' *Alpine Journal*, vol. 41 (1929), pp. 88–98.

SMYTHE, F. S. 'Wind Slab on Kangchenjunga.' *British Ski Year Book*, vol. 5 (1930), p. 375.

SOMIS, I. *A true and particular account of the most surprising preservation and happy deliverance of three women buried 37 days by a heavy fall of snow at the village of Bergemoletto in Italy* (London, 1765).

SPRECHER, F. 'Alpiner Skilauf und Lawinengefahr.' *Jahrbuch des Schweiz.* Skiverbandes, 1912.

SPRECHER, F. 'Grundlawinenstudien 1.' *Jahrbuch Schweiz. Alpenclub*, vol. 35 (1899–1900), pp. 268–292.

SPRECHER, F. 'Grundlawinenstudien 2.' *Jahrbuch Schweiz. Alpenclub*, vol. 37 (1901–1902), pp. 219–243.

SPRECHER, F. 'Ueber die künstliche Veranlassung des Abganges von Lawinen.' *Schweiz. Zeitschrift für Forstwesen*, vol. 61, 1910.

ST. LOUP. *La montagne n'a pas voulu* (Paris, 1949).

STRABO. *Strabo's Geography* (translation by Jones, London & New York, 1917–32).

STREIFF, R. 'Schnee und Lawinen.' *Die Alpen*, vol. 12 (1936), pp. 46–57.

SWISS FED. INST. FOR SNOW AND AVALANCHE RESEARCH. *Lawinenverbau im Anbruchgebiet* (Official Swiss guidelines, Bern, 1961). English translation: *Avalanche Control in the Starting Zone* (Colorado State Univ., Fort Collins, 1962).

SWISS FED. INST. FOR SNOW AND AVALANCHE RESEARCH. *Winterberichten*, Nos. 1–42 inclusive (1936/37–1976/77).

SWISS RED CROSS. *Rechenschaftsbericht über die Lawinenkatastrophe* 1951 (Bern, 1953).

TYLER, J. E. *The Alpine Passes—962–1250 A.D.* (London, 1930).

TYNDALL, J. *Hours of Exercise in the Alps* (London, 1871).

U.S. DEPT. OF AGRICULTURE FOREST SERVICE. *Snow Avalanches—a handbook of forecasting and control measures* (Washington, 1961).

U.S. DEPT. OF AGRICULTURE FOREST SERVICE: PERLA R., MARTINELLI, M. JR. *Avalanche Handbook* (Agriculture Handbook 489. Washington 1976).

VOELLMY, A. 'Über die Zerstörungskraft von Lawinen.' *Schweiz. Bauzeitung*, vol. 73 (1955).

WAGNER, A. 'Luftbewegung bei Lawinenstürzen.' Pp. 126–131 in *Praktische Schnee und Lawinenkunde* by W. Paulcke (Berlin, 1938).

WECHSBERG, J. *Avalanche!* (London, 1958).

WINTERHALTER, R. U. 'Vom Schnee.' *Die Alpen*, vol. 31 (1955), No. 2.

WINTHROP YOUNG, G. *Mountain Craft* (London, 1945).

ZDARSKY, M. *Beiträge zur Lawinenkunde* (Vienna, 1929).

ZDARSKY, M. *Elemente der Lawinenkunde* (Vienna, 1916).

ZINGG, TH. 'Wetter und Lawinenkatastrophen.' *Die Alpen*, vol. 31 (1955), No. 2.

ZSIGMONDY/PAULCKE. *Die Gefahren der Alpen* (Munich, 1911).

Index